Proceedings of the National Conference
on

Future Electricity Systems
Challenges and Current Trends
(NCEFES 2021)

Organized by

Institution of Engineers (Jodhpur Local Center)
under the aegis of Electrical Division Board

ALLIED PUBLISHERS PVT. LTD.

New Delhi • Mumbai • Kolkata • Chennai • Bangalore • Hyderabad

ALLIED PUBLISHERS PRIVATE LIMITED

D-5, Sector-2, **Noida**–201 301
Ph. Nos.: 0120-4320295/2542557/4352866 • E-mail: delhi.books@alliedpublishers.com

17 Chittaranjan Avenue, **Kolkata**–700072
Ph.: 033-22129618 • E-mail: cal.books@alliedpublishers.com

15 J.N. Heredia Marg, Ballard Estate, **Mumbai**–400001
Ph.: 022-42126969 • E-mail: mumbai.books@alliedpublishers.com

No. 25/10, Commander-in-Chief Road, Ethiraj Lane (Next to Post Office)
Egmore, **Chennai**–600008
Ph.: 044-28223938 • E-mail: chennai.books@alliedpublishers.com

P.B. No. 9932, No. 15, 3rd Floor (Next to Vijaya Bank), 5th Cross, Gandhinagar, Karnataka, **Bangalore**–560009, Ph.: 080-41530285 / 22386239
E-mail: bngl.journals@alliedpublishers.com / apsabng@airtelmail.in

Sri Jayalakshmi Nilayam, No. 3-4-510, 3rd Floor (Above More Super Market)
Barkatpura, **Hyderabad**–500027
Ph.: 040-27551811, 040-27551812 • E-mail: hyd.books@alliedpublishers.com

Website: www.alliedpublishers.com

ISBN: 978-93-90951-33-8

Published by Sunil Sachdev and printed by Ravi Sachdev at Allied Publishers Pvt. Ltd., D-5, Sector-2, Noida–201 301

Proceedings of the National Conference
on

Future Electricity Systems
Challenges and Current Trends
(NCEFES 2021)

Foreword

Dear Authors, Esteemed Readers,

It is with insight satisfaction that we write this Foreword to the book that features selected papers from the 36th National Convection of Electrical Engineers and Conference held in hybrid mode and organized by Institution of Engineers Jodhpur Local Centre, Jodhpur, India on November 27–28, 2021.

The conference particularly encouraged the interaction of research scholars, academicians and industry persons with established community to discuss recent developments in electrical technology for sustainable smart cities and energy management. Their contributions helped to make the conference as outstanding as it has been.

In addition to the contributed papers, Prof. M.S. Thackar Memorial Lecture has been delivered by Prof. Jagdish Kumar, Head, Electrical Engineering Department, Punjab Engineering College, Chandigarh. Moreover, Prof. Deepak Fulwani, IIT Jodhpur, Prof. Rohit Bhakar, MNIT Jaipur and Prof. Dinesh Birla, Rajasthan Technical University, Kota delivered the keynote speeches.

We are thankful to the track chairs and reviewers for extending their support in various forms to make a sound technical program. We are grateful to the members of the advisory committee for their timely suggestions and guidance. We sincerely appreciate and acknowledge the wholehearted support extended by the members of the organizing and technical committee.

We thank all authors and participants for their contributions. Last, but surely not the least, we thank Mr Jagdish Singh and his team of Allied Publishers for rendering their excellent support in bringing out this proceeding.

We hope that this book will prove to be an excellent reference material for researchers and academician working in the field of future electricity systems and allied areas.

Akhil Ranjan Garg
Om Prakash Mahela
Baseem Khan
Neeraj Gupta
Emarti Kumari

Preface

Currently, electrical power distribution systems have been shifting from traditionally designed system (high voltage level to lower voltage level) to smart grids (small generations are connected to the distribution networks) and green energy sources (wind energy, solar energy). The efficiency of existing traditional energy network is poor, because it suffers from various failures and uncertainties. The renewable sources are limited, therefore, the future electricity systems requires massive development of distributed energy resources (DER), especially distributed generation (DG) based on renewable energy resources (RES). The present book includes the selected papers of the National Conference on Future Electricity Systems: Challenges and Current Trends (NCEFES 2021), which held on November 27–28, 2021, in Jodhpur, India. The NCEFES 2021 as a national academic event on future electricity systems. It provided a premier platform for electrical engineers, researchers, practitioners and educators to present and discuss the challenges, current trends and concerns as well as practical challenges encountered and solutions for future electricity in the fields of **Future Electricity Systems**. The book features original papers presented by graduate students, research scholars, academicians, and industry persons during this conference. The topics covered in the book include recent advances in Distributed generation and Power quality, Optimization techniques, Renewable energy/ Alternative energy, Reliability of distributed energy systems, Smart microgrid, Advanced monitoring and Novel control strategies, Real-time simulation and contingencies analysis, Ancillary services and metering, Economic benefits, Application of machine learning, Data acquisition, Internet of Things (IOT), Load forecasting, Future electricity systems, Integration of communication technology, Blockchain technology and its application in Energy systems, Cloud computing for energy, Cyber physical energy systems, Renewable energy grid integration, Smart protection techniques for Electrical distribution network, Recent developments in electrical technology for sustainable smart cities, Energy management.

Prof. Akhil Ranjan Garg
Conference Chair
Jai Narain Vyas University, Jodhpur

Structure of the Book

The book is organised into 23 chapters based on quality researcher papers accepted after rigorous review of submitted papers in National Conference on **Future Electricity Systems**:

Chapter 1 provides a review of electrification and power system flexibility to achieve net-zero carbon emission. Carbon neutrality can be acquired by promoting renewable energy sources to generate electricity, sector coupling to generate green hydrogen, and electrification of sectors like transportation, gas, and heating. Grid flexibility can accommodate these challenges of net load demand and have the potential to handle the peak demand of the system.

Chapter 2 deals with different prospects of electrode material selection of ultra-capacitor for energy storage systems. The critical components of ultra-capacitor are electrode, electrolyte and separator; also discussed in detail that specific area on the electrode surface will be enhanced by incorporating nanomaterial such as CNT, graphene and activated carbon (AC).

In **chapter 3** discussed about the artificial neural network strategy for a Li-ion battery management system with the estimation of state-of-charge and the charging-discharging behaviour of a Lithium-ion battery. The study of energy management for hybrid energy storage system connected to PV system is elaborated in **Chapter 4**. This strategy is based on monitoring of generation-demand mismatch and divides into low frequency and high frequency terms. High frequency terms are handled by SC and low frequency terms are handled by battery.

Chapter 5 deals with power consumption optimization techniques in base station systems. The effect of temperature and irradiance on grid connected solar PV systems have discussed in **Chapter 6.** 120 KW grid-tied solar PV array is simulated in MATLAB and performance evaluation of the system conducted. Electric process heater and heater control methodology are demonstrated in **Chapter 7** using thyristors and contactors or combination of both.

Chapter 8 discussed the synchronized detection and discrimination of external and internal shunt faults in fixed series compensated transmission line (FSCTL) using two zones current information based on wavelet analysis (WA). Application based rapid prototype has been developed and elaborated in **Chapter 9.** A hardware was developed for rapid real time electrical applications using TMS320F28335 DSP and MATLAB.

Chapter 10 explained the cooling methods for solar PV panels under extreme loading in summers such as 60°–70°C specifically in west Rajasthan. Also discussed a systematic approach for the selection of appropriate PCM. **Chapter 11** deals with the bioreactor performance under the influence of time delays. Performance improvement of MPPT based grid connected hybrid renewable energy system is discussed in **Chapter 12**. MPPT use increases conductivity technology and in photovoltaic and wind power systems. In this chapter also carried out the detailed analysis of solar, wind and PMSG modelling.

In **Chapter 13** discussed the particle swarm optimization (PSO) technique based on swarm intelligence. The optimal PID controller design has been implemented using MATLAB/Simulink and compared the classical tuning method with PSO based PID controller. A review of flexible generator planning of power system is described in **Chapter 14**. Controlling and monitoring of feeders in power distribution system using supervisory control and data acquisition (SCADA) system is explained in **Chapter 15**. SCADA systems are widely used to monitor and remotely control the distribution of power to all substations. **Chapter 16** deals with harmonic analysis and mitigation techniques for modern electrical devices such as Compact Fluorescent Lamp (CFL). Also discussed about development of Simulink models of the CFL circuits with the filters on LT spice software.

In **Chapter 17** evaluated the techno-economic feasibility of the system and introduced a grid-connected solar/wind/biogas hybrid renewable energy sources (HRES). Also calculated the power requirement of Electrical Engineering department, CTE Udaipur Rajasthan to meet their demand. Mathematical modelling of photovoltaic system in MATLAB is elaborated in **Chapter 18**. A system in which PV model is used to charge battery of electric vehicle using Single Ended Primary Inductor Converter has been discussed in **Chapter 19** to reduce the model transportation cost. MPPT and P&O algorithms were demonstrated to design the SEPIC converted for the maximum power of solar panel.

Chapter 20 deals with evaluation of grid connected wind farm with variation in wind speed. How to effect on the stability of DFIG by the fluctuations of the wind speed? Discussed in this chapter in detail. Machine learning technique to predict the IMDb rating of movies has been discussed in **Chapter 21**. Hollywood movies list was collected from Wikipedia and IMDb and developed a real-world data set to predict the rating of up-coming movies. In **Chapter 22** applied the fractional order controller to enhance the performance of Electric vehicle. Also compared the PID, FOPID controller mechanisms and various optimization techniques generate better output. In **Chapter 23** discussed the existing SCADA system and compared it with Phasor Measurement Units. Also suggested some corrections to redefine the existing SCADA system.

Committees

National Advisory Committee

Patron Prof (Dr) Santanu Chaudhury, Director,
Indian Institute of Technology, Jodhpur

Chairman Er Narendra Singh, FIE, President, IEI

Co-Chairman Mr P. Rajamani, FIE, Chairman, ELDB, IEI

Convenor Prof (Dr) S.K. Calla, FIE, Member, ELDB, IEI

Co-Convenor Prof (Dr) Avdhesh Sharma, FIE, Committee Member, IEI, JLC

Members

Mr Pankaj Dadwal, FIE, Member, ELDB, IEI
Prof Himanshu Bikas Goswami, FIE, Member, ELDB, IEI
Mr Mahendra Ratanlal Kothari, FIE, Member, ELDB, IEI
Dr Natarajan Kumaresan, FIE, Member, ELDB, IEI
Mr Rup Lal Mahajan, FIE, Member, ELDB, IEI
Mr M. Marbaniang, FIE, Member, ELDB, IEI
Mr Nagamaheshwar Rao Nadella, FIE, Member, ELDB, IEI
Mr K.K. Gopalakrishnan Nair, FIE, Member, ELDB, IEI
Mr P.R. Natarajan, FIE, Member, ELDB, IEI
Prof (Dr) Khiroda Kumar Rout, FIE, Member, ELDB, IEI
Mr Ashwani Kumar Sharma, FIE, Member, ELDB, IEI
Mr Donray Awungshi Shishak, FIE, Member, ELDB, IEI
Mr Nishi Kant Prasad Sinha, FIE, Member, ELDB, IEI
Mr R.R. Tanwar, FIE, Member, ELDB, IEI
Dr B. Venkataramana, FIE, Member, ELDB, IEI
Mr Ravinder Kumar Sharma, FIE, Member, ELDB, IEI
Dr S. Vasantharathna, CIT, Coimbatore
Dr E. Chandirasekaran, CIT, Coimbatore
Dr Alok Kanti Deb, IIT Kharagpur
Dr Rajeev Gupta, RTU, Kota
Dr Jayashri Vajpai, MBM Jodhpur
Dr D.K. Palwalia, RTU, Kota
Dr Dinesh Birla, RTU, Kota

Technical Committee

Publication Committee

Organizing Committee

Chairman Dr Akhil Ranjan Garg, FIE, Chairman, IEI, JLC

Co-Chairman Dr Ghanshyam Vaishnav, Secretary, IEI, JLC

Organising Secretary Dr Kailash Choudhary, MIE, MBM Jodhpur

Members

Er Daria Singh Yadav, MIE, IEI, JLC, Jodhpur
Er Santosh Kumari Meena, EED, MBM Jodhpur
Er Amit Meena, MBM Jodhpur
Dr Sanjay BC Gaur, JIET Jodhpur
Er Chandrashekhar Singh, JIET, Jodhpur
Er Khushboo Shah, EED, MBM Jodhpur
Er Danveer Rajpal, EED, MBM Jodhpur
Er Harish Khayani, EED, MBM Jodhpur
Er Avnish Bora JIET Jodhpur
Er A.K. Singh, MIE, IEI, JLC, Jodhpur
Dr Emarti Bhaskar, MBM Jodhpur
Er Abhishek Gaur, MBM Jodhpur

About the Editors

Akhil Ranjan Garg

Dr. Akhil Ranjan Garg is presently working as Professor and Head in the Department of Electrical Engineering, Faculty of Engineering and Architecture, J.N.V. University Jodhpur since February 2011. Prof. Garg had served as visiting Faculty at IIT Jodhpur from July 2010 to December 2018. Dr. Garg completed his B.E. (Electrical Engineering), M.E. (Control Systems) from M.B.M. Engineering College (FOEA, J.N.V.U. Jodhpur) and subsequently did his Ph.D. from IIT Delhi. He is having more than 28 years of teaching and research experience since his appointment as Assistant Professor in year 1992. Dr. Garg has made pioneering research contributions and has solved many open problems in research areas that include Computational Neuroscience, Intelligent Systems, Pattern Recognition, Power Electronics, Power Quality, Grid integration of Renewable Systems and Machine Learning. He has published over 55 papers in these areas in refereed international journals and conference proceedings.

Dr. Garg is a recipient of numerous honors and awards including DAAD (German Academic Exchange Program) Fellowship, US Naval Academy Fellowship, IBRO (International Brain Research Organization) Fellowship. Dr. Garg is Fellow Institution of Engineers (India), Life Member Indian Society of Technical Education (ISTE), Member IEEE, Member International Brain Research Organisation (IBRO) and Senior Member International Neural Network Society (INNS).

Om Prakash Mahela

Dr. Om Prakash Mahela (Senior Member, IEEE) received the B.E. degree from the College of Technology and Engineering, Udaipur, India, in 2002, the M.Tech. degree from Jagannath University, Jaipur, India, in 2013, the Ph.D. degree from IIT Jodhpur, India, in 2018, all in electrical engineering, and the M.B.A. degree in human resource management from Indira Gandhi National Open University, New Delhi, India, in 2021. From 2002 to 2004, he was an Assistant Professor with the Rajasthan Institute of Engineering and Technology, Jaipur. From 2004 to 2014, he was a Junior Engineer with Rajasthan Rajya Vidyut Prasaran Nigam Ltd., India, and an Assistant Engineer, since February 2014. He has authored more than 175 research articles and book chapters. He has edited two books. He performed more than 220 reviews for the prestigious journals of IEEE, Elsevier, Springer, Willey, and Taylor & Francis.

His research interests include power quality, power system planning, grid integration of renewable energy sources, FACTS devices, transmission line protection, and condition monitoring. He was a recipient of the University Rank Certificate, in 2002; the Gold Medal, in 2013; the Best Research Paper Award, in 2018; and the C.V. Raman Gold Medal, in 2019. He received the certificates of outstanding contribution in the reviewing from *Computers and Electrical Engineering*, *International Journal of Electrical Power and Energy Systems*, *Measurement*, and *Renewable and Sustainable Energy Reviews*.

Research Profiles

- Scopus: https://www.scopus.com/authid/detail.uri?authorId=56045098300
- Web of Science Researcher ID: O-7868-2017, https://publons.com/researcher/1221860/om-rakash-mahela/
- Author ORCID ID=56045098300, http://orcid.org/0000-0001-5995-6806
- Google Scholar: https://scholar.google.com/citations?user=rz69hqoAAAAJ&hl=en&oi=ao
- Research Gate: https://www.researchgate.net/profile/Om_Mahela2?ev=hdr_xprf

Research Specializations

Power Quality, Power System Protection, Reliability, Power System Planning, Grid Integration of Renewable Energy

Baseem Khan

Dr. Baseem Khan (M'16) received his Bachelor of Engineering degree in Electrical Engineering from Rajiv Gandhi Technological University, Bhopal, India in 2008. He received his Master of Technology and Doctor of Philosophy degrees in Electrical Engineering from the Maulana Azad National Institute of Technology, Bhopal, India, in 2010 and 2014, respectively. Currently, he is working as faculty member at Hawassa University, Ethiopia. His research interest includes power system restructuring, power system planning, smart grid technologies, meta-heuristic optimization techniques, reliability analysis of renewable energy system, power quality analysis and renewable energy. He has to his credit more than 50 research publications in international Journals and conference proceedings.

Research Profiles

- Google Scholar: https://scholar.google.com/citations?user=WAPWo64AAAAJ&hl=en
- Scopus: https://www.scopus.com/authid/detail.uri?authorId=55647924400
- LinkedIn: www.linkedin.com/in/baseem-khan-01ab73125
- Research Gate: https://www.researchgate.net/profile/Baseem_Khan2
- Publons: https://publons.com/researcher/1356176/baseem-khan
- KUDOS: https://growkudos.com/profile/baseem__khan

Neeraj Gupta

Dr. Neeraj Gupta is a Teaching & Research Staff in the Dept of Computer Science and Engineering at Oakland University, Rochester, Michigan. Before that, he served as visiting Asst. Professor in the same department from 2017 October. In addition to the academic services, Dr. Gupta is a consultant and Sr. Scientist in MIS 2000 Inc/Global Electronics, Nevada, USA. His research career started in 2013 after getting Ph. D. from the Indian Institute of Technology (IIT) Kanpur, where he joined the IIT Jodhpur, India as a teaching and research staff. Thereafter, he started an international research carrier as an Assistant. Professor at University for Information Science and Technology, "St. Paul the Apostle", Ohrid, Macedonia until 2017. Due to the exposition of different engineering fields and wide research domain, his current research interests are in the field of evolutionary Deep learning, Edge enabled deep learning solutions, Evolutionary computation, Optimization, Smart grid, Smart data, Multi-agent modelling, and Smart agriculture. Smart vehicles, Cyber security, Biometrics, Federated learning IoT and applications, development of heuristic optimization algorithms, particularly in the area of multilateral and real-time operation of the complex systems. Currently, he has one patent on the Deep AI Engine, which develops highly optimized deep learning models for edge devices, automatically, without expert intervention and for any type of data.

Emarti Kumari

Dr. Emarti Kumari is presently working as Assistant Professor in the Department of Mechanical Engineering, Faculty of Engineering and Architecture, J.N.V. University Jodhpur since September 2017. Dr. Kumari completed his B.E. in Mechanical Engineering from M.B.M. Engineering College, FOEA, J.N.V.U. Jodhpur, India in 2004. She received M. Tech. (Design Engineering) and Ph.D. (Solid Mechanics) from IIT Delhi, in 2010 and 2019, respectively. Her research interest includes finite element methods, plates and shells, laminated composite materials, bending, buckling and vibration analysis of thin-walled structures, non-linear static and dynamic analysis of wind turbines. She has published over 14 papers in these areas in refereed international journals and conference proceedings.

Dr. Emarti is Life Member Institution of Engineers (India), Member American Society of Mechanical Engineers (ASME), Member Indian Society of Applied Mechanics (ISAM).

Research Profiles

- Scopus: author Id= 57193837339
- Web of Science Researcher ID: ABD-7740-2020
- http://orcid.org/0000-0002-3339-8153
- Google Scholar: https://scholar.google.com/citations?user=7JNRQ0gAAAAJ&hl=en&oi=ao
- Research Gate: https://www.researchgate.net/profile/Emarti-Kumari

The Institution of Engineers (India)

AN ISO 9001 : 2015 CERTIFIED ORGANISATION

(ESTABLISHED 1920, INCORPORATED BY ROYAL CHARTER 1935)

8 GOKHALE ROAD, KOLKATA – 700020

A Century of Service to the Nation

Mr P. Rajamani, *FIE*
Chairman
Electrical Engineering Division Board
The Institution of Engineers (India)

Chairman's Message

It is a great pleasure to note that the Jodhpur Local Centre of The Institution of Engineers (India) has come out with this proceeding of the 36th National Convention of Electrical Engineers which was organized on the theme **Future Electricity Systems: Challenges and Current Trends**.

With growing concern about the climate change and global warming people around the world are becoming conscious about the environmental impact of the generation of electricity through conventional method using fossil fuels. The primary challenge in the near future is to supply uninterrupted, cheap and clean electricity to meet the ever increasing demand. The only viable solution is to exploit the potential of renewable energy sources primarily the solar and wind and to some extend small hydro generator.

However, this reorientation of the energy sources will result in small generating unit connected to large network contrary to the traditional concept of distribution of power from large generating sources to the utility. A development of active distribution network management, from centralized to more distributed system management, is needed. Information, communication, and control infrastructures will be needed with increasing complexity of system management.

Telephone: (033) 40106299; Web: http://www.ieindia.org

All these issues were deliberated during the two days National Convention and the ideas and solutions are collated into these beautiful proceedings. I hope this proceeding will be very helpful for the professional engineers.

Thank you.

P. Rajamani

The Institution of Engineers (India)

(Established 1920, Incorporated by Royal Charter 1935)
8 Gokhale Road, Kolkata–700020

Prof (Dr) Swapan Bhaumik, *FIE*
Chairman,
Committee for Advancement of
Technology and Engineering

Mobile: 7005032903
E-mail:
drsbhaumik@gmail.com

Chairman's Message

I am happy to note that the Jodhpur Local Centre of The Institution of Engineers (India) is organinsing the 36th National Convention of Electrical Engineers on the theme **"Future Electricity Systems: Challenges and Current Trends".**

As the future of the Electricity System will drastically change over next few years globally due to surge in the demand for clean and green electricity which will have minimum impact on the environment. Almost all the countries will give up on conventional source of energy production and will embrace non-conventional source of electricity. The whole system of electricity management starting from generation to distribution will have to be remodeled. The economics associated with the trade will also see a radical change.

Considering the above the theme selected by the Jodhpur Local Centre is very appropriate. It is expected that the proceedings of this Convention being published will contain large numbers of quality papers which will be helpful to the researchers, academicians, students and professional engineers.

I wish Jodhpur Local Centre all the best and hope a grand success for the convention.

Thank you

Prof (Dr) Swapan Bhaumik

Office & Residence
Professor and Former Head, Mechanical Engineering Department,
National Institute of Technology, Agartala–799046
'Matrikalaya', Road No.16, PO Dhaleswar, Agartala–799007, Ph: 0381-2559119

Contents

Electrification and Power System Flexibility: A Review

Abhishek Harit*, Prerna Jain and Rohit Bhakar
Department of Electrical Engineering,
Malaviya National Institute of Technology,
JLN Marg, Jaipur-302017, India
E-mail: *2020ree9541@mnit.ac.in

ABSTRACT: *In COP26 meet, Supreme leaders of almost 200 countries are committing to achieve net-zero by a defined time. Carbon neutrality can be acquired by promoting renewable energy sources to generate electricity, sector coupling to generate green hydrogen, and electrification of sectors like transportation, gas, and heating. All of these solutions to reduce carbon emission are proficient to change the scenario of the peak load of any country and its net load characteristics. Grid Flexibility can accommodate these challenges of net load demand and have the potential to handle the peak demand of the system. A review on Electrification and power system flexibility is presented to understand the issues of power system flexibility in the evolving environment of future electrification.*

Keywords: Power System Flexibility; Electrification; Renewable Energy Sources; Demand Side Flexibility.

Introduction

Entire globe is concerning about the issues of climate change and the rising temperature of earth. Carbon emissions through various sources are considered as a prime culprit for the Global Warming. One of the proven solution to reduce the carbon emission is using of renewable sources to generate electricity and making them as major source of generation. Restructuring of the power system also introduces the high penetration of renewable sources in the power system by allowing distributed energy resources (DER) at the demand side. DERs are available in various forms, generation through a roof top solar panel over a residential or industrial premises, electric vehicle (EV), micro grids, smart mini grids etc. Despite of numerous advantages, high integration of renewable energy sources in the power system creates the issues variability and uncertainty of the supply. This intermittent nature of renewable sources can jeopardize the reliability and security of the system which is a prime concern of the power system operator. Power system flexibility can be a key solution to handle this variation of the supply.

Ref. [1] gives a fundamental insight to the concept of power system flexibility and its various resources available in the system. Having resources is not just sufficient its optimal usage is also important to optimize the resource, and power system reserve is also discussed thoroughly. Ref. [4] discussed about various challenges, evaluation parameters of power system flexibility and impacts of RES on power system transient and small signal stability of the system. Implementation of resources from local system to system wide is also presented.

Global Economies of the world are also emphasizing electrification of the sector like transportation, gas network and heat network, which could also evolve new business model, market design and encourages the consumers for being a part of the electricity market [9, 10]. Heating sector will create more impact on demand as compared to transportation [14]. This globalisation has change the definition and role of consumers to convert them from passive consumers to active consumers called prosumers. Chief Technology Officer (CTO) of a pioneer company states that carbon-neutral energy system can be achieved by electrifying the system, which works much more flexible and interconnected power system [15, 16]. This paper presented a review on the electrification and its impact on power system flexibility along with the new challenges.

The structure of the paper is given as below Power system flexibility, Characteristics of Flexible resources, Assessment of Power System Flexibility, Electrification and its Challenges of Power System Flexibility, Conclusion followed by References.

Power System Flexibility

Power system flexibility is defined as the capability of a power system to cope with the variability and uncertainty that VRE generation introduces into the system in different time scales, from the very short to the long term, avoiding curtailment of VRE and reliably supplying all the demanded energy to customers [2]. Ref. [2] presented the evolution and necessary criteria for the definition of power system flexibility from different institutions and researchers.

Power System Flexibility can be provided through different facets of the system like Generation side, Network side, Storage side, Market side and Demand side [1, 2]. Flexibility resources can also categorize on the basis of different indices which are ramping limit, power capacity, energy capacity and time response. Ramping limit denotes the capability of source to provide power per minute (MW/min) to reach its maximum value, Power capacity shows the capability of providing maximum power (Pmax in MW), Energy capacity denotes the capability of providing maximum power for a certain time period (MWh), and Time response is the ability of the flexibility resource to respond. Concept of power system reserve for the flexibility to accommodate the uncertainty issues of RES in operation and planning is also discussed [1].

Different facets of flexibility and their resources are: [4]

1. Generation side Flexibility can be attain through the fast operating power plants also called dispatchable power plants e.g. hydro power station, open cycle gas turbine, closed cycle gas turbine, load following power plants. Conventional power plants having the issues like low ramping rates, minimum up and down time can also be retrofitted for the fast operation, spinning reserve for scheduling, managing the regulation of generators.
2. Network side Flexibility can be achieved by grid strengthening and expansion, by using FACTs devices, cross border interconnection and dynamic line rating.
3. Storage side Flexibility includes sector coupling technology like power to gas (hydrogen), power to heat, battery storage system, electric vehicle, pumped hydro storage plants.
4. Market side Flexibility can be accommodated at least cost which is associated by operation of the electricity market, advanced forecasting to get less errors, time granularity, space granularity, economic dispatch, unit scheduling, support from ancillary services.
5. Demand side flexibility can be attain by demand response program by using TOU (time of use) tariff, ancillary services through an aggregator, large consumers can directly provide the flexibility to Distribution system operator.

Ref. [2] discussed classification of power system flexibility resources and proposed comprehensive classification. Ref. [3, 5] discussed the flexibility classification on the basis of different time scale of short, medium and long term and also proposed a methodology to evaluate the flexibility requirement for annual, weekly, daily time frame. A case study is done for French power system for market potential of flexibility solutions and policy decision for the implications.

Characteristics of Flexibility Resources [2]

Researchers have defined characteristics of flexibility resources on the basis of different focus area, like focussing on the participation of resources in the wholesale market, three commonly used set for the characteristics of flexibility resources are power capacity (MW), ramp rate (MW/min) and ramp duration (MWh). Another framework includes two aspects; technical characteristics, economical characteristics. Technical characteristics are further classified as quantitative, qualitative and control technical characteristics, whether classification of economical aspect are given as capital economic characteristics (CAPEX) and operational economic characteristics (OPEX). These characteristics can be summarize as:

A. *Quantitative Technical Characteristics:* It is defined as the capability of flexibility resources expressed numerically with defined units. All the technical parameters of any flexibility resource comes under this category like power capacity, ramping capacity, energy capacity, ramp duration, service duration, response time, rebound effect, recovery time, ramp frequency, flexibility time, minimum up/down time, responsiveness (in %), efficiency, energy loss.
B. *Qualitative Technical Characteristics:* It is defined as the quality of the flexibility resources expressed in degree of comparison. These characteristics contain location, predictability, credibility, and ownership.
C. *Control Technical Characteristics:* It is defined as, how the flexibility resources are controlled enough to provide flexibility. This can be subcategorized into direct control (explicit control) and indirect control (implicit control).
D. *Capital Economic Characteristics (CAPEX):* It includes necessary investments in flexibility resources itself and investment costs related to enabling activation of flexibility.
E. *Operational Economic Characteristics (OPEX):* It includes costs related to activation, cycling costs and penalty for non-delivery.

Assessment of Power System Flexibility

Rich literature is available for the evaluation of power system flexibility. Researchers uses flexibility indices as well as the metrics to get judgement and measure the required flexibility in the system. Some literature also recommends tools like; GIVAR (Grid Integration of Variable Resources),

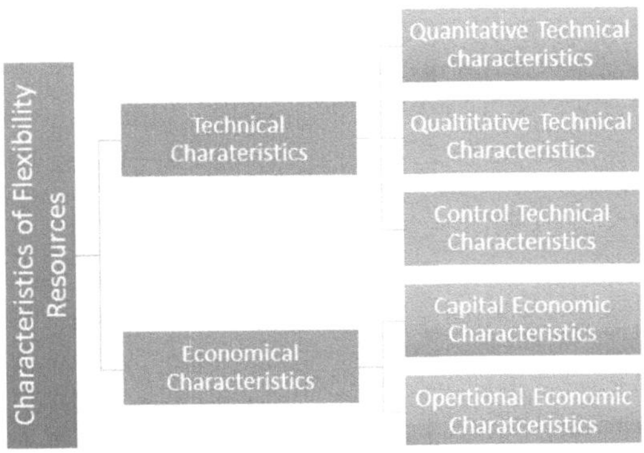

Fig. 1: Characteristics of Flexibility Resources

FAST2 (Flexible Assessment Tool) for quick assessment of required flexibility. Ref. [6] proposed a model for the quantification of integrated steam and energy system, applied over two industries to use industrial steam on turbine generator to produce electricity and provide it as ancillary services. Ref. [7] proposes frequency spectrum based tools flexibility solution modulation stack (FSMS) and flexibility solution contribution distribution (FSCD) for the flexibility quantification in different time frames. FSMS derives how resource provide flexibility and FSCD evaluates the contribution to flexibility. These tools can be derived from former available data and simulation data of the power system. In the literature, researchers also proposed methods to uncertain net load to measure the flexibility.

Ref. [8] discussed methods to measure net load uncertainty and flexibility requirements; Exogenous approach and Endogenous approach are explained with their limitations that both approaches do not calculate the feasibility and optimality of flexibility simultaneously because in the uncertain set the feasibility can be found but due to the shifting of operation point it does not guarantee an economic solution. A new framework is given, so that both the things can be done simultaneously. Two stage robust optimization, in which maximum variation range of uncertainty that flexible resource can provide without compromising the reliability is obtained and then feasible uncertainty set is taken as an input set and worst case of recourse in minimized.

Electrification and its Challenges of Power System Flexibility

Electricity becomes a continuously evolving sector. United States and European Countries are the most active economies in this evolution. These countries are also growing fast in the direction of the electrification of the sectors specifically transportation and heating. Sectors which aren't electrified directly, uses the resources from electrification indirectly like aviation or marine sector, they can be operate on green hydrogen produced by renewable energy sources. With electrification, prosumers become increasingly dynamically and participate in the operation of the power system.

A report by NREL analysed the scenario of electrification in whole US and their potential impacts on the system. By modeling overview of the demand side scenario and demand side flexibility (DSF) helps the system to achieve the operational benefits by providing two types of services: energy shifting and operating reserves. Energy shifting is nothing but the changing of consumption pattern (high consumption during low demand and vice versa), while in the operating reserve DSF reduces the need for other generators such as natural gas plants and storage to provide reserves. Analysis results that DSF can be attained by 71% of the operating reserve which is greater than storage reserve (63%) and concludes that operational efficiency of such systems can be enhanced through the expansion of demand-side flexibility, especially flexibility from newly electrified loads [11, 12].

European Union established an internal energy market as Energy Balancing (EB) markets for more competitiveness and flexibility in the interconnected network of the continent. Transmission system operators (TSO) have to implement four platforms for energy balancing; Exchange of balancing energy from Replacement Reserve (RR), Exchange of balancing energy from manual frequency restoration reserve (mFRR), Exchange of balancing energy from automatic frequency restoration reserve (aFRR), Exchange of balancing energy from Imbalancing Netting (IN). TSOs have established projects to implement this exchange of energy balance [17]. European smart grid task force identified barriers on the deployment of demand side flexibility with the various perspectives and also proposed recommendations to enhance the demand response to overcome barrier gaps. Table 1 summarizes the barriers for different stances and recommendation at country level or European level [18].

Table 1: Barriers and Recommendations for Demand Side Flexibility [18]

Perspectives	*Barriers*	*Recommendations*
General Technical Issues	• Smart meter usage without consumer permission • Strategy for whole EU • Insufficient data of load and generation forecasting at Demand side • Inadequate grid codes in power system • Less Observability in low voltage power systems	• Increasing low voltage observability by data acquired from smart meters • Improve forecasting techniques for DER at distribution level • Grid codes shall be assessed to ensure behaviour of demand response technology in power system • Aware consumers about smart meter, data security, and inform consumer about data sharing protocol
Consumer Scenario	• Need for Standardisation • Interoperability • Consumer awareness and security • Demand side Incentives	• Develop framework of standardisation and interoperability of hardware like charging station, smart meters • Information, awareness and trust would be develop by the coordination of stakeholders • New offers shall be developed to encourage new service providers
Market Access and Coordination	• Framework for Demand Side Response providers • Integration of Implicit and Explicit DR • Data security, access and Sharing • Market Accessibility • Market framework and market efficiency	• Platform for data acquisition from consumers' smart meter • Framework of demand side response need to be develop • Securing data sharing and improve cyber security • New market model framework including demand side response required to activate DSF with transparency need to be developed

(Contd.)

(Table 1 contd.)

Perspectives	*Barriers*	*Recommendations*
Product Design	• Importance of Location based Information • Prequalification of assets • Information of products • Multiple balance responsible parties at single connection point	• Sharing the information of location needs to be mandatory part for congestion management products • System operator or market operator must define the qualification of the prosumers to provide service • Demand side flexibility must be provided by smart meter data and telemetry requirements should be established to collect data and execution of new product design

An analysis has been conducted by IRENA over 30 innovative methods under 4 categories; Enabling Technologies, Market Design, Business Models and System Operation to make the system more flexible to achieve electrification in holistic way through the possible solutions, execution and costing issues of these innovative methods. This analysis leads towards the potential of flexibility providing by different methods, their infrastructural and implementation cost and challenges to implement that methods are carried out. Outcome of the analysis can be seen by the Figures 2 and 3 which shows the relation between level of flexibility required, costing range of the methods and level of complexity [19]. This report concludes that countries at starting level of the implementation of grid flexibility are encouraged to focus on demand side management and market design innovations i.e. moderate investment for moderate potential of flexibility whether advanced countries must promote the mini grids, smart grids innovation, storage side flexibility, power to X innovations i.e. high investments to attain flexibility of high impact on the system [19].

Various challenges to implement flexibility in the electrification are:

A. *Regulatory Challenges:* Upgradation of regulation and policy for small term and long term contracts, investing and incentivizing schemes for prosumers or DSOs are required for energy transition, new business model and new market design to drive the power system smoothly, strong regulations for cross border power transfer are also required to avoid miscommunication [24].

B. *Investment and Infrastructure Development*: As electrification extends rapidly in the countries like USA, UK, Germany etc. new infrastructure is required to be develop to increase generation capacity, transmission network capacity and distribution network capacity to deal with bi direction flow of power such as smart meters and mini grids, handling of peak demand due to electrification also needed to be taken care off, electrification also encourage the distributed energy resources to participate in the system as demand side flexible resources,

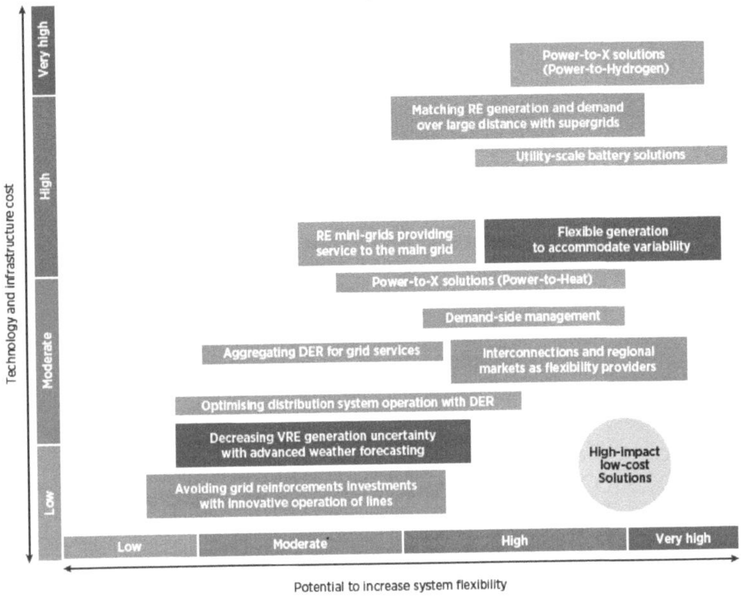

Fig. 2: Costing of Flexibility *vs.* Potential to Provide Flexibility [19]

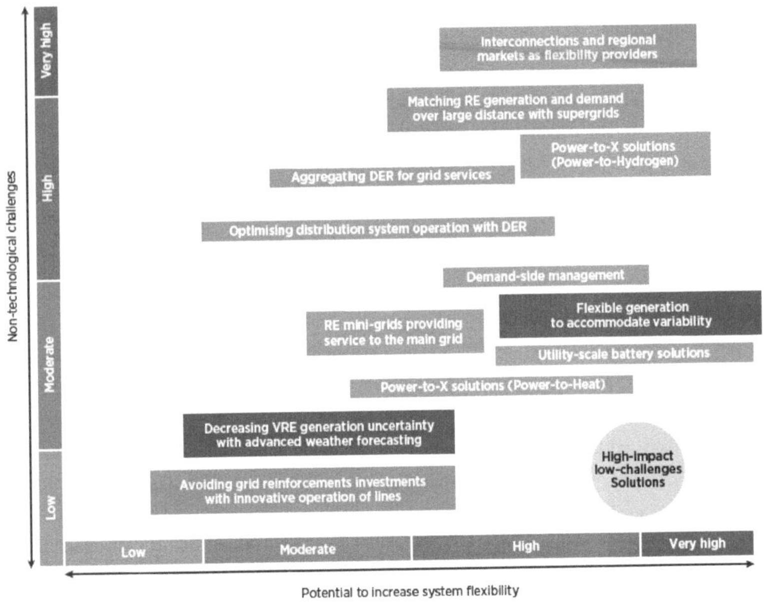

Fig. 3: Challenges in Implementation of Flexibility *vs.* Potential to Provide Flexibility [19]

which can possibly change the roles and responsibilities of the stakeholders (DSO-TSO), investment and management for storage system at utility level are the one of the primary challenge [11, 12, 24].

C. *Stakeholders Communication and Coordination Challenge:* As demand side resources increasing complexity in the network, challenges of congestion and co-ordination also arises for stake-holders (DSO-TSO-Aggregators), contract settlement, procuring services at local level, energy balancing role because prosumer can sell their service to either DSO or TSO so proper coordination is needed between them is one of a great challenge [21, 24]. A price signal based approach by system operators (TSO and DSO at their levels) is proposed to minimise the electricity cost to the consumer [23].
D. *Multi Energy System:* Single market seems great for the overall flexibility of all the sectors but there are lots of challenges at execution level to handle complex system with three different market (electricity, gas and heating) at same platform [18, 22]. Ref. [20] discussed the Multi Energy vectors in a multicarrier methodology for end users.

CONCLUSION

Carbon neutrality is strongly associated with the electrification of the region, which requires large investment and development of infrastructure to in-crease the capacity of generation trans-mission and distribution. Electrification through renewable energy sources is a challenging task which requires deep analysis, rigorous planning and high level of flexibility to handle the mismatch of supply and demand. This has changed the scenario of demand side dynamically and encourage the passive consumers to become prosu-mers. It also impacts the load curve and peak demand of the power system and shift the paradigm of power system flexibility in two ways:

- Increase in demand requires flexibility to handle the balance of supply and demand to ensure about the frequency in the safe operational range.
- Large number of consumers with stored energy try to provide service to get incentivized from the energy market, either through direct contract or through an aggregator.

Both these scenarios vitalize the weightage of demand side flexibility in overall flexibility. Country like USA and European Countries after analysing the future scenario of electrification, have turned their focus towards the demand side flexibility and started working on the solution like energy balancing markets, regulatory policies, market designs, business models, operation of market from DSO pers-pective, improvement in TSO-DSO coordination.

No one-size-fits-all because every country or region has different geographical conditions, different dominating renewable resources, different investing plans, and strategies but some key innovations that can be applied and implemented effectively to any region are advanced forecasting to handle the uncertainty of VRE, demand-side management, market improvement, coordination of stakeholders and change in policy regulation of the power markets.

References

[1] Mohandes, B., Moursi, M.S.E., Hatziargyriou, N. and Khatib, S.E., "A review of power system flexibility with high penetration of renewables" *IEEE Transaction on Power Systems*, Vol. 34, 2019.

[2] Degefa, M.Z., Sperstad, I.B. and Saele, H., "Comprehensive classifications and characterizations of power system flexibility resources", *Electric Power System Research,* Vol. 194, pp. 107022, May 2021.

[3] Akrami, A., Doostizadeh, M. and Aminifar, F., "Power system flexibility: an overview of emergence to evolution", *Journal of Modern Power Systems and Clean Energy., IEEE*, Vol. 7, pp. 987–1007.

[4] Impram, S., Nese, S.V. and Oral, B., "Challenges of renewable energy penetration on power system flexibility: A survey" *Energy Strategy Review,* Vol. 31, pp. 100539, September 2020.

[5] Heggarty, T., Bourmaud, J.Y., Girad, R. and Kariniotakis, G., "Multi temporal assessment of power system flexibility requirement" *Applied Energy,* Vol. 238, pp. 1327–1336, March 2019.

[6] Xu, X., Sun, W., Abeysekera, M. and Qadrdan, M., "Quantifying the flexibility from industrial steam systems for supporting the power grid" *IEEE Transactions on Power Systems*, Vol. 36, 2021.

[7] Heggarty, T., Bourmaud, J.Y., Girad, R. and Kariniotakis, G., "Quantifying power system flexibility provision", *Applied Energy,* Vol. 279, 2020.

[8] Pourahmadi, F., Hosseini, S.H. and Firuzabad, M.H., "Economically optimal uncertainty set characterization for power system operation flexibility" IEEE Transactions on Industrial Informatics, Vol. 15, 2019.

[9] Hermans, M., Bruninx, K. and Delarue, E., "Impact of CCGT start up flexibility and cycling costs toward renewables integration", *IEEE Transactions on Sustainable Energy*, Vol. 9, 2018.

[10] Moghaddam, I., Chowdhury, B. and Doostan, M., "Optimal sizing and operation of battery energy storage systems connected to wind farms participating in electricity markets", *IEEE Transactions on Sustainable Energy*, Vol. 10, 2019.

[11] Zhou, E. and Mai, T., "Electrification Futures study: Operational analysis of US power systems with increased electrification and demand side flexibility", *National Renewable Energy Laboratory*, May 2021.

[12] Murphy, C. *et al.*, "Electrification futures study: Scenarios of power system evolution and Infrastructure development for the US", *National Renewable Energy Laboratory*, Jan. 2021.

[13] Belhomme, R., Corsetti, E., Gutschi, C. and Kessels, K., "Bottom-up flexibility in multi energy systems" *IEEE PES Magazine,* Vol. 19, July/August 2021.

[14] Wang, S., Tan, X., Liu, T. and Tsang, D.H.K., "Aggregation of demand side flexibility in electricity markets: negative impact analysis and mitigation method" *IEEE Transactions on Smart Grid*, Vol. 12, 2021.

[15] "Future energy scenario", *National grid Electricity System Operator,* July 2021.

[16] "Accelerated grid development is indispensable for speeding up the European energy transition", *Perspectives, Hitachi Energy,* March 2021.

[17] "ENTSO-E Balancing report", published by European network of transmission system operator for electricity, 2020.

[18] "Deployment of demand side flexibility: Perceived barriers and proposed recommendations" *Expert Group 3, European Smart Grid Task Force,* 2019.

[19] "Innovation landscape for a renewable powered future: Solution to integrate variable renewables", *International Renewable Energy Agency,* 2020.

[20] Malley, M.J. *et al.*, "Multicarrier energy systems: Shaping our energy future", *Proceedings of the IEEE,* Vol. 108, No. 9, 2020.

[21] Khajeh, H., Firoozi, H., Hesamzadeh, M.R., Laaksonen, H. and Shafie-Khah, M., "A local capacity market providing local and system wide flexibility services", *IEEE Access,* Vol. 9, 2021.

[22] Orths, A. *et al.*, "Flexibility from energy systems integration", *IEEE PES Magazine*, Vol. 19, November/December 2019.

[23] Zotti, G., Pourmousavi, S., Morales, J., Madsen, H. and Poulsen, K., "A control based method to meet TSO and DSO ancillary services needs by flexible end users", *IEEE Transactions on Power Systems*, Vol. 35, May 2020.

[24] Baker, P., "Challenges facing distribution system operator in a decarbonised power system", *Regulatory Assistance Project, Belgium,* 2020.

Different Prospects of Electrode Material Selection of Ultracapacitor based Energy Storage System

Aditya Kachhwaha* and Ashish Godara

Department of Electrical Engineering, M.B.M. Engineering College, J.N.V. University, Jodhpur, Rajasthan–342011, India

E-mail: *Aditya.k1692@gmail.com

ABSTRACT: *Ultracapacitor are being accepted as an electrical energy storage (EES) device for wide range of electrical power applications. The energy from the non-conventional plants cannot be fully realized without EES. The solution of this problem is that ultracapacitors should be used either as primary power source or a hybrid source as in combination with batteries or other sources. Ultracapacitor can fill an important space in the current state of ESS. The ultracapacitor's key capability of high power density (KW/Kg) is mainly governed by its high value of faradic capacity range, which further depends upon the characteristics of electrode materials. The new material such as graphene-carbon nanotube (CNT) composites is used for higher faradic capacity which has 3-D structure with exceptional properties such as superior electrical conductivity and an extensive surface area better than that of alone carbon nanotubes (CNTs) and graphene materials respectively. In this manuscript, factors affecting the characteristic enhancement and classification of ultracapacitor based on the electrode material selection is discussed.*

Keywords: Ultracapacitor, Supercapacitor, Graphene-carbon Nanotubes (CNT) Composite, Electrolytic Double Layer Capacitor (ELDC).

Introduction

The world's energy demand is increasing continuous due to rising in population and the improvement of the living standards. This increases level of the pollution and causes the need of non-conventional sources [1]. Since, the non- conventional electrical power cannot be generated continuously which causes the imbalance of energy requirement, this will disturb the quality and stability of the electrical power supply. So, for an alternative solution to meet up the energy demand electrical energy storage (EES) is required [2, 3]. Ultracapacitors and batteries are the two main types of EES devices. Ultracapacitors are EES devices that is used to store and release charge at very fast rate. The requirement of ultracapacitor in the field of non- conventional energy storage is emerged as they have high specific power, long cycle life, less charging, better portability and discharging time (approx. 100 times less than the battery), temperature range is –40°C to 60°C and has higher power capability of charging and discharging in a very short time duration, which makes it fit for energy storing application in case of renewable sources [2]. The high faradic capacity of an ultracapacitor is basically subjected to the electrode material performance. The main drawback of the ultracapacitor is their low operation voltage (near 2.5 volts), which is not advisable in high voltage applications [3]. Ultracapacitor intended for use in higher power application [5], should have high value of faradic capacitance and cell voltage. So there is a need of improvement in ultracapacitor by using different combinations of electrode material and electrolyte type. This paper discusses the recent advancement in the field of ultracapacitor and comparative study of different types of ultracapacitors based on its electrodes materials.

Ultracapacitor Working

There are three main storage principles on which electrodes of the ultracapacitor works, which contribute to the capacitance of device, stated as follows:

- Electrolyte Double-layer capacitor is a type which is using electrostatic phenomenon for storing electrical energy.
- Pseudocapacitance which is using the electrochemical storage of the electrical energy (using redox reactions with charge-transfer) [22].
- By using Hybrid storage technique.

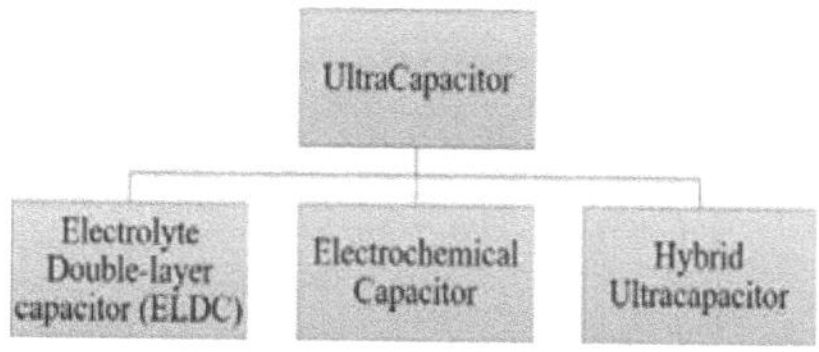

Fig. 1: Basic Classification of Ultracapacitor

Electrolytic Double Layer Capacitor (EDLC)

EDLC is called "electrical double layer capacitor," because a charge layers of ions is made up between the electrolyte and electrodes that is used in it as the dielectric material between them [6], [7]. EDLC consist of two similar electrodes made up of highly porous material like activated carbon or charcoal. These materials possess high surface area (1500–2000 m^2g^{-1}), good electronic conductivity and high electrochemical stability. Charge storage takes place by reversible adsorption of the ions at the surface of active material. The large number of ions can penetrate in the surface of electrode. The space between the two activated carbon electrodes is filled with the solution of electrolyte which is also separated by a substrate electrode material called as a dielectric separator that provide insulation to the electrodes. The Voltage of EDLC becomes low with discharging and increases in case of charging, as seen from the Figure 2. Actual EDLC capacitance value can be measured but it varies and depends on charge and discharge characteristic condition [8].

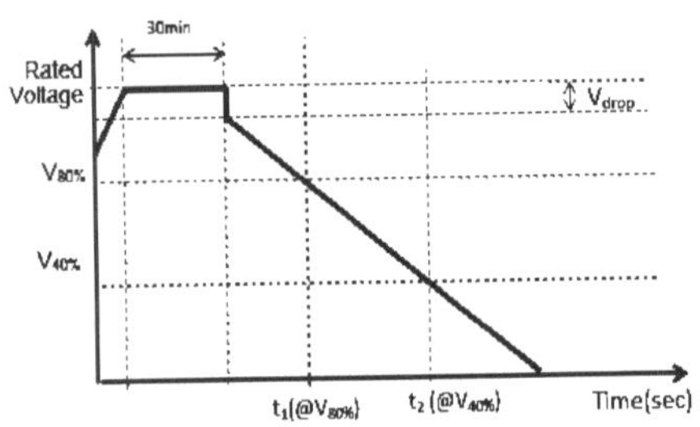

Fig. 2: ELDC Terminal Charatreristic Curve

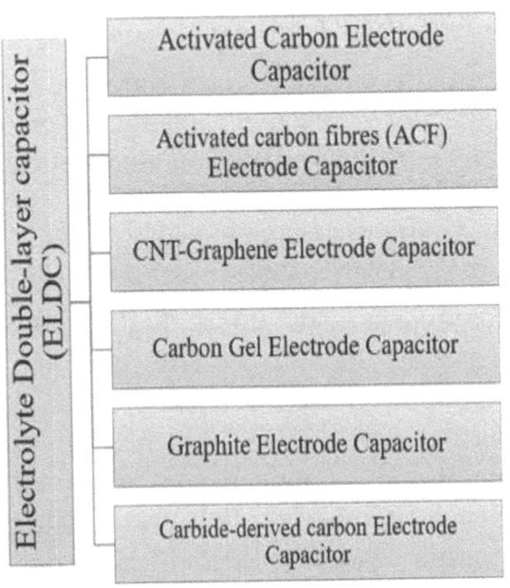

Fig. 3: Different Types of Electrolytic Double Layer Capacitor

Activated carbon fibers (ACF) are derived from the activated carbon. They have 10 μm diameter and micropores with a very narrow pore-size distribution. The AFC surface area is about 2500 m²/g [9–11]. A highly porous, synthetic and light material Carbon aerogel is obtained from an organic gel in which its liquid of gel is replaced with a gas known as "frozen smoke". Carbon aerogel electrode has 104 F/cm³ volumetric capacitance and about 400–1200 m²/g gravimetric densities, that is capable of retrieving 325 kJ/kg of energy density and power density of 20 W/g [12–14]. Carbide-derived carbon (CDC), which is also termed as a tunable nanoporous carbon, generally obtained from carbide precursors, such as titanium carbide and binary silicon carbide. Energy density of CDC supercapacitor is 10.1 Wh/kg, Faradic capacitance is 3.5 kF and also have over 106 charge/discharge cycles [15].

Electrochemical Ultracapacitor

Electrochemical capacitance also termed as a Pseudocapacitance, It is the property of the pseudocapacitor which uses the electrochemical storage of electricity, in this type of capacitor (Ultracapacitor) charge transfer is facilitated by a very fast type of sequential reversible faradic redox reactions and also by the process of intercalation on the electrodes surface [11], [12], [16], [17]. So that it is responsible for charge transfer between electrolyte and electrode coming from a redistribution of ions. There are two electrode materials are mainly used to store charge in pseudocapacitors as follows in Figure 4.

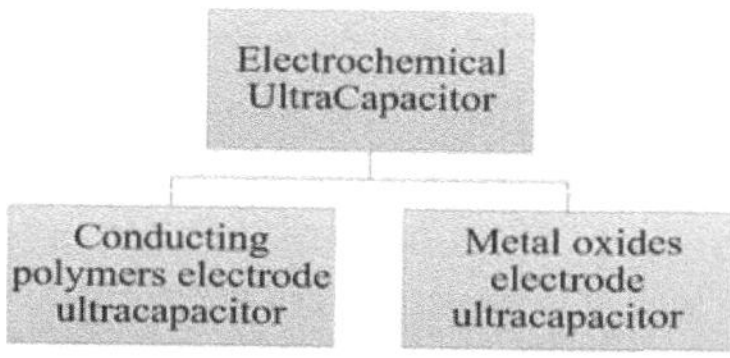

Fig. 4: Electrochemical Capacitor Types

Conducting Polymers

They are basically electron-conducting polymers but have characteristic of pseudo capacitive material. Mechanically weak conducting polymers have high conductivity which results in a low equivalent series resistance (ESR) and also have high faradic capacitance value. Conducting polymers with above characteristics are polypyrrol, polythiophene, polyaniline, and polyacetylene. The problem of limited cycling stability is associated with the conducting polymer electrodes. Although polyacene conducting polymer electrodes can provide up to 10,000 cycles [18]–[20].

Metal Oxides

This type includes the Oxides of transition metals like as iridium (IrO_2), ruthenium (RuO_2), manganese (MnO_2), iron (Fe_3O_4), and sulphides such as titanium sulphide (TiS_2) alone or in combination with electrolyte of H_2SO_4 can give 720 F/g specific capacitance and the energy density up to 26.7 Wh/kg (96.12 kJ/kg) [16]. Since there is a voltage dependency of the charge transfer by these reactions, that's why a capacitive phenomenon occurs in it. Two types of charge transfer voltage dependent reactions are involve in it. First are redox reaction and another is reaction corresponding to adsorption of ions [7], [21], [22].

Hybrid Ultracapacitor

Ultracapacitors that can be made up with one electrode of pseudo-capacitance material and the other electrode of double-layer carbon material are referred as hybrid capacitors [23]–[25]. Most of the hybrid capacitors are developed currently have used nickel oxide as positive electrode of the pseudo-capacitance material [1].

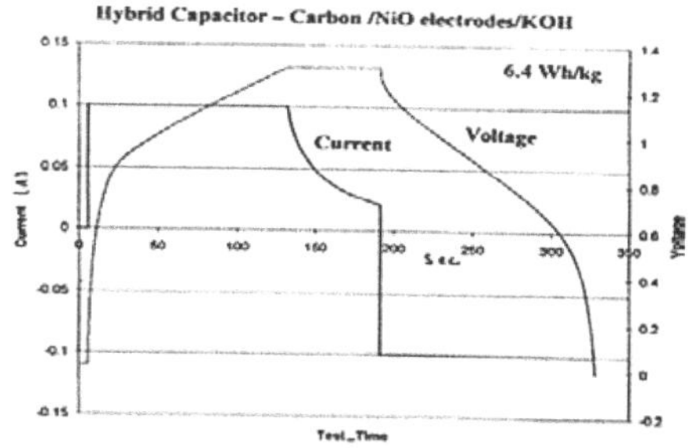

Fig. 5: Voltage *vs.* Charge of Hybrid Ultracapacitor

In hybrid ultracapacitor the energy density is significantly higher than the other type of ultracapacitors. Its Voltage *vs.* Charge graph is nonlinear and non-ideal as shown in Figure 5. Hybrid capacitors can be made of using two non-similar doped conducting polymer materials and also by mixed metal oxide.

The hybrid capacitors can be classified by their electrode configuration as: composite, asymmetric, and battery-type respectively [24].

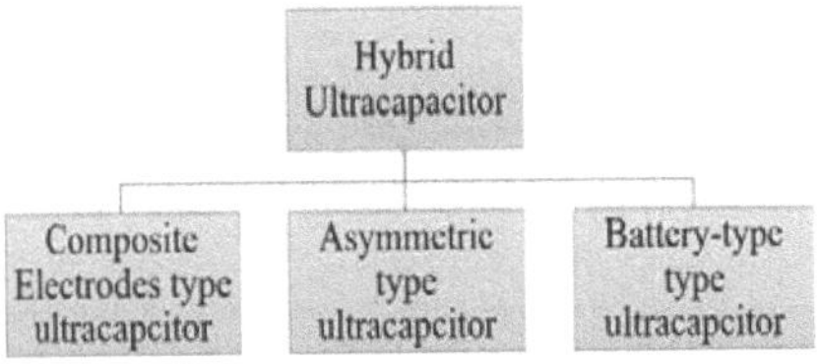

Fig. 6: Hybrid Capacitors can be Classification

Composite Electrodes

They are constructed from carbon polypyrrole a conducting polymer and nanotubes. The carbon based materials which are highly porous used to create a capacitive double-layer so that they have a high-surface-area that increases penetration of ions between electrolyte and the deposited pseudo capacitive materials.

Asymmetric

It combines both faradic and non-faradic processes by using two different electrodes, one EDLC electrode with pseudo capacitor electrode. Thus by this configuration higher specific energy up to 36–72 kJ/kg and power density up to 10–20 Wh/kg as well as higher voltage can be achieved [8].

Battery-type

It also couples two different electrodes like asymmetric type but in this case one electrode from battery type is coupled with the ultracapacitor electrode. Thus the higher energy demand can be fulfilled by this configuration of ultracapacitor and higher power batteries and by combining the energy characteristics of both, batteries with the high cycle life & power, and short period charging time of ultracapacitor. Suitable materials which are used in this type of configuration are mainly LTO ($Li_4Ti_5O_{12}$), nickel hydroxide and lead dioxide as one electrode and activated carbon as the other.

Fundamental Factor for Ultracapacitor Upgrading

There are two main key factors to improve the performance of the ultra-capacitors:

1. Increasing the surface area of the coating of electrode plate by turning it to highly porous.
2. By increasing the maximum voltage that the ultracapacitor can handle without breakdown.

Capacitance in farads is measure of how much electrical energy can be stored in particular ultracapacitor so that it will hold a certain amount of voltage. Now if the voltage is increased then the amount of energy increases too. Some key points are as follows regarding the advancement of ultra-capacitor:

- For higher capacitive rating, there is a need of highly porous activated material so that more and more electron can penetrated in it. Graphene has the capability to boost the surface area. Carbon nanotubes (CNT) and graphene can also withstand the higher voltage than that of the charcoal material (activated carbon).
- Another points regarding advancement in ultracapacitor is the improving voltage strength. For

this there should be improvement in dielectric electrolyte solution. The ionic liquid electrolytes (all ion, no solvent, behaves like a liquid) can be operated at voltage more than that of conventional electrolytes. But ionic liquids tend to freeze below room temperature which are also expensive, and having higher resistance than conventional electrolytes.

Electrode Material Used in Ultracapacitor

The main factors of affecting the performance of ultracapacitors are the capacity of the electrodes to store and release energy that is always determined with the electrode material and effective penetration of electrons. This is responsible for the determination of breakdown potential strength of the ultracapacitor. Now a day's ultracapacitor with graphene based electrode is used for higher power rating application. The active material loading and electrode thickness is also the key factor in electrode performance as the higher energy density can be achieved with a thicker electrode and similarly the higher power density can be achieved with a thinner electrode. Therefore, it is very important to have electrode with comparable thickness to achieve accurate and desired rating of ultracapacitor [9].

Advancement Using Graphene Material in Ultracapacitor

Although, activated carbons are used commercially in the supercapacitors as electrode materials, there still exists a need to increase the energy density without compensating for high power density and cycle life. Activated carbons have disordered micropores, with a broad pore sizes distribution whereas microporous carbons made using templating method are better for use as high energy density electrode materials, this is due to a narrow pore size distribution [5], [19]. It has 2-D structure which gives it properties such as enhancement in surface area than the carbon nanotubes (CNTs) and superior electrical conductivity. Graphene is the best option for ultracapacitor application because its surface area doesn't vary due to distribution of the pore size and also provides electrolyte access to both of its surfaces. Graphene has a theoretical specific surface area of 2630 m^2/g which can theoretically lead to a capacitance of 550 F/g and by using ionic electrolytes at voltages up to 4 V. A specific energy density of graphene is found as 85.6 Wh/kg (308 kJ/kg) [7], [14]. The size depends upon quantization effect in semiconductor and their optical properties of nano crystals which are defined by their size and surface area. Nanomaterials have highly exotic properties and the properties of nanomaterial's are:

- High mechanical strength in micro level.
- High surface energy, large specific heat capacity and surface to volume ratio are also high.
- Large thermal expansion.
- Highly affected from catalytic activity and have high level of self-diffusion.

Magnetic susceptibility and sintering rate are high. These properties of nano-particles are useful in effectively penetration of electrons in electrode material, thus the faradic capacity of ultracapacitor increases [7]. These modifications is done for improving the capacitance rating of ultracapacitor and capacitance is increasing due to increase in the effective surface area. Using modified graphene material with fewer layers, so there would be a low aggregation on the surface area and surface area will increase significantly by this configuration of graphene. It leads to a higher faradic capacitance. Structure of graphene layers is as shown in following Figure 7 [7], [8].

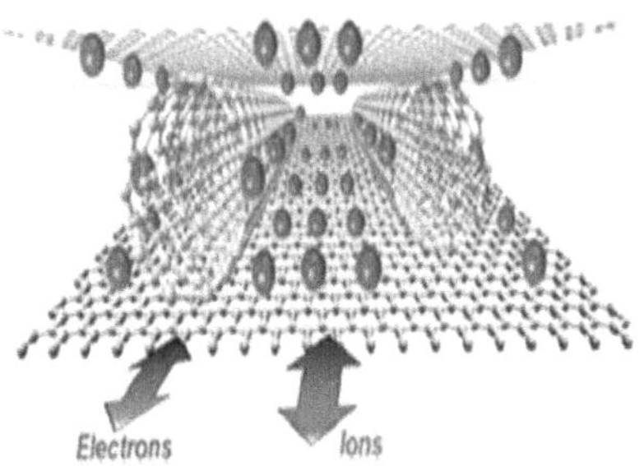

Fig. 7: Modified Graphene Material with Multi-Layers

A graphene based advance composites involves CNT materials that are inserted between the graphene sheets [24]. Graphene offers more specific area (2630 m^2/g) and CNT act like spacers as a conducting path to enable absorption of a larger numbers of ions on graphene surface.

With this composites material combination as electrode, high energy density (62.8 Wh/Kg) and power density (58.5 KW/Kg) can be obtained with organic electrolytes. The energy density can be further increased up to 155 Wh/Kg by incorporating this composite with ionic electrolyte.

Table 1: Advance Carbon Electrode Materials

Types	*V (Volt)*	*KW/ Kg*	*Wh/ Kg*
PbO_2 (1M/H_2SO_4)	2.25	15.7	8.9
$Li_4Ti_5O_{12}$ Graphite + Organic	3.7	50	7
$Li_4Ti_5O_{12}$ AC Carbon + Organic	2.8	18	4.5
AC Carbon + Organic	2.7	5.7	6.4
AC Carbon + (1M/H_2SO_4)	1	1.7	1.2

In case of pseudocapacitors, aggregation of graphene nanosheets limit the capacitance of graphene and graphene derivatives to 100–200 F/g. Polyaniline (PANI) is a famous conducting polymer with high conductivity, low cost, high flexibility and multiple-redox state. It has high pseudo capacitance but low stability during charge/discharge.

The composites of PANI and graphene can overcome these drawbacks. Graphene/PANI nanofibers have faradic capacity of 210 F/g and 194 F/g at 0.3 Ag and 3 Ag respectively. rGO/ PANI composites have the faradic capacity of 320 F/g at 0.1 Ag. The utilization of electrode material can be increased by using ordered PANI nanowires. Graphene oxide (GO) sheets are used for preparation of PANI arrays, based on van der Waals force [20].

Importance of Separator Material in Ultracapacitor

Separator is basically a Porous paper, polymer and ceramic that is used in ultracapacitor to provide insulation between electrodes. The features of separator are as follows:

- It is responsible for ionic conduction between electrodes.
- It also offers resistance to components which is proportional to its thickness and inversely proportional to its area.
- Its conductivity is proportional to its Porosity which is like 50–70%.
- High power operation of ultracapacitor depends on its low internal resistance.

The most commonly used separator materials are polymer membranes, cellulose, and glass fibres. The selection of separator materials is influenced by the nature of electrode, voltage rating and temperature.

Electrolyte Solutions

Electrolytes used in the ultracapacitor can be of different types and the key parameters which depends on it are energy density and power density. Both are affected by the types of electrolytes.

1. The energy density is responsible for the voltage window and ion concentration of the electrolyte in ultracapacitor.
2. The power density is mainly affected by the resistance of the electrolyte.

Following type of electrolytes are generally used in ultracapacitors.

(a) Aqueous electrolytes
(b) Organic electrolytes and
(c) Ionic liquid electrolytes

The choice of electrolytes used in ultracapacitor are mainly depends on the following parameters:

- Resistance and Capacitance offered by solution.
- Ease of production of material.
- Potential range (in which the electrolyte will remain stable)

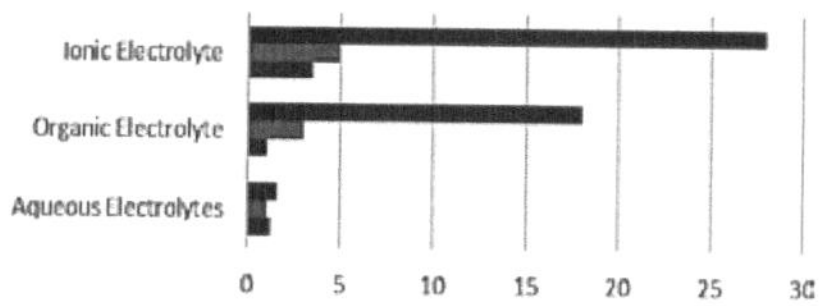

	Aqueous Electrolytes	Organic Electrolyte	Ionic Electrolyte
■ Resistivity (ohm-cm)	1.6	18	28
■ Cell Voltage Range (volts)	1	3	5
■ Density (g/cc)	1.25	1	3.5

Fig. 8: Comparision amongst Different Electrolytes

Conclusions

An overview of energy storage system especially for the ultracapacitor energy storage system is discussed in this paper. The critical components of ultracapacitor are electrode, electrolyte and separator. It is concluded that by incorporating nanomaterials like CNT, graphene, AC (Activated Carbon) etc. specific area on the electrode surface is increased significantly which further directly related to increment in faradic rating. The CNT-graphene composites have ability to improve ultracapacitor characteristic parameters such as potential window (up to 3V), energy density and power density. Further higher increment in faradic rating and energy density CNT-Graphene is used with ionic electrolyte. Consequently the main field of research for enhancement of ultracapacitor characteristic parameters are based on CNT-graphene multi-layer composite electrode materials immersed in ionic liquid electrolytes.

References

[1] Burke, A., "Ultracapacitors: why, how, and where is the technology", *Journal of power sources*, Vol. 91, pp. 37–50, 2000.

[2] Obreja, V.V.N., "On the Performance of Commercial Supercapacitors as Storage Devices for Renewable Electrical Energy Sources".

[3] Koohi-Fayegh, S. and Rosen, M.A., "A review of energy storage types, applications and recent developments", *Journal of Energy Storage*, Vol. 27, 2020.

[4] Sahay, Kuldeep and Dwivedi, Bharti, "Supercapacitors Energy Storage System for Power Quality Improvement", *An Overview J. Electrical Systems* (2009).

[5] Kachhwaha, A., Shah, V.A. and Shimin, V.V., "Integration methodology of ultracapacitor-battery based hybrid energy storage system for electrical vehicle power management," *2016 IEEE 7th Power India International Conference (PIICON)*, Bikaner, 2016, pp. 1–6.

[6] Sharma, S., Panwar, A.K. and Tripathi, M.M., Storage technologies for electric vehicles, *Journal of Traffic and Transportation Engineering,* 2020.

[7] Afif, Ahmed; Rahman, Sheikh M.H.; Azad, Atia Tasfiah; Zaini, Juliana, Islan, Md Aminul and Azad, Abul Kalam, "Advanced materials and technologies for hybrid supercapacitors for energy storage—A review", *Journal of Energy Storage,* Vol. 25, 2019.

[8] Signorelli, Riccardo; Ku, Daniel C.; John G. Kassakian, Life and Schindall, Joel E., "Electrochemical Double-Layer Capacitors Using Carbon Nanotube Electrode Structures" *IEEE,* Vol. 97, No. 11, November 2009.

[9] Volfkovich, Yu. M., Mikhailin, A.A., Bograchev, D.A., Sosenkin, V.E. and Bagotsky, V.S., Studies of Supercapacitor Carbon Electrodes with High Pseudocapacitance, A.N. Frumkin Institute of Physical Chemistry and Electrochemistry, Russian Academy of Sciences, Moscow, Russia, Dr. Ujjal Kumar Sur (Ed.), ISBN 978-953-307- 830-4

[10] Jackel, N., Weingarth, D., Zeiger, M., Aslan, M., Grobelsek, I. and Presser, V., "Comparison of carbon onions and carbon blacks as conductive additives for carbon supercapacitors in organic electrolytes", *Journal of Power Sources,* 272 (2014), 1122–1133.

[11] Vivekchand, S.R.C.; Rout, Chandra Sekhar; Subrahmanyam, K.S., Govindaraj, A. and Rao, C.N.R., "Graphene-based electrochemical supercapacitors", *J. Chem. Sci.*, Vol. 120, No. 1, January 2008, pp. 9–13. © Indian Academy of Sciences.

[12] Volfkovich, Yu. M., Mikhailin, A.A., Bograchev, D.A., Sosenkin, V.E. and Bagotsky, V.S., "Studies of Supercapacitor Carbon Electrodes with High Pseudocapacitance", *Recent Trend in Electrochemical Science and Technology*.

[13] Pandolfo, A.G. and Hollenkamp, A.F., "Carbon properties and their role in supercapacitors", *Journal of Power Sources,* 157 (2006), 11–27.

[14] Liu, Chenguan; yu, Zhenning; Neff, David; Zhamu, Aruna and Jang, Bor Z., "Graphene based supercapacitor with an ultrahigh Energy density".

[15] SkelCap High Energy Ultracapacitors—Data Sheet (PDF) Skeleton Technologies. Retrieved 2015-02-28.

[16] Zheng, J.P., Cygan, P.J. and Jow, T.R., "Hydrous Ruthenium Oxide as an Electrode Material for Electrochemical Capacitors".

[17] Mahulkar, Niraj and Karandikar, P.B., "Comparison of Different Types of Carbon Materials as Electrodes of Super Capacitor", *International Journal of Innovative Research in Science, Engineering and Technology*, Vol. 3, Issue 6, June 2014.

[18] Gao, Lijun, Dougal, Roger A. and Liu, Shengyi, "Power Enhancement of an Actively Controlled Battery/ Ultracapacitor Hybrid", *IEEE transactions on power electronics,* Vol. 20, No. 1, January 2005.

[19] Zhang, Li Li and Zhao, X.S., "Carbon-based materials as supercapacitor electrodes". *Chem. Soc. Rev.,* 2009, 38, 2520–2531.

[20] Wang, Li; Ye, Yinjian; Lu, Xingping; Wen, Zhubiao; Li, Zhuang; Hou, Haoqing and Song, Yonghai, "Hierarchical Nanocomposites of Polyaniline Nanowire Arrays on Reduced Graphene Oxide Sheets for Supercapacitors".

[21] Mohapatra, S., Acharya, A., Roy, G.S., "The role of nanomaterial for the design of supercapacitor", *Lat. Am. J. Phys. Educ.*, Vol. 6, No. 3, Sept. 2012.

[22] Tan, Yu Bin and Lee, Jong-Min, "Graphene for supercapacitor applications", *Chem. A*, 2013, 1, 14814–14843.

[23] Chenguang, L., Zhenning, Y., Neff, D., Zhamu, A. and Jang, B.Z. (November 2010). "Graphene-based supercapacitor with an ultrahigh energy density". *Nano Lett.,* 10(12): 4863–4868.

[24] Raghavendra, Kummara Venkata Guru; Vinoth, Rajangam; Zeb, Kamran; Chandu V.V., Muralee Gopi; Sambasivam, Sangaraju; Kummara, Madhusudana Rao; Obaidat, Ihab M. and Kim, Hee Je, "An intuitive review of supercapacitors with recent progress and novel device applications", *Journal of Energy Storage,* Vol. 31, 2020,

[25] Kowsari, Elaheh, "High-Performance Supercapacitors Based on Ionic Liquids and a Graphene Nanostructure".

Artificial Neural Network Strategy for a Li-ion Battery Management System

Ankush Koli[1], Hemani Paliwal[2] and Vikramaditya Dave[3]

Department of Electrical Engineering, College of Technology and Engineering, MPUAT, Udaipur, Rajasthan–313001, India

E-mail: [1]ankushshakya13@gmail.com; [2]hemani.8588@gmail.com; [3]vdaditya1000@gmail.com

ABSTRACT: *In this paper, Artificial Intelligence introduces to the artificial neural network strategy for estimation of state-of-charge and the charging-discharging behavior of a Lithium-ion battery management system used for supplying electricity to an electric motor in the electric vehicles. Battery management system is viable for prediction of available capacity in the battery. Besides the Coulomb Counting and Kalman Filters methods of SoC estimation, artificial neural network gives an accurate estimation of the state-of-charge with efficient operation of battery. Experimental methods of SoC estimation can be regulated in labs to analyze the behavior of the battery whereas the battery based ANN model uses training data to preserve the health of the battery with improving the characteristics of the battery. Experimental method is compared with the trained data of an ANN model and the result can be obtained as an accurate estimation of SoC and health of the battery.*

Keywords: Electric Vehicle (EV), Battery Management System (BMS), State-of-Charge (SoC), Artificial Neural Network (ANN).

Introduction

Due to the increasing rate of usage of internal combustion (IC) engine based conventional vehicles and the raising prices of the fossil fuel used in it, global warming and green-house gases are rising rapidly. So, electrification of the vehicles is an actual need to reduce temperature of the earth and carbon-dioxide emission. Now-a-days, electric vehicles contain promising importance to reduce global warming and provide zero emission to reduce air pollution and improve the air quality of the environment. Battery is a key part of the EVs for supplying electric power to an electric motor. A Lithium-ion battery popularly used to supply electric power to an induction motor used in this work. Battery of an EV can be charged by plugging the vehicle into a power source or under the operation of regenerative braking mode.

Battery management system (BMS) is engaged with a Lithium-ion battery to maintain the cells in a battery for proper operation of the EVs [1]. BMS provides a solution for battery modeling, state estimation and charging/discharging behaviour including monitoring component. BMS introduces to the state-of-charge estimation relative to the capacity of the battery used in the EVs.

Basic structure of a Battery Management System is shown in the Figure 1. A battery management system:

(a) protects the electric cells of a battery from overheating and overcharging.
(b) protects the battery pack from damaging.
(c) increases the life cycle of a battery.
(d) records the current and voltage parameters of a battery during charging and discharging.

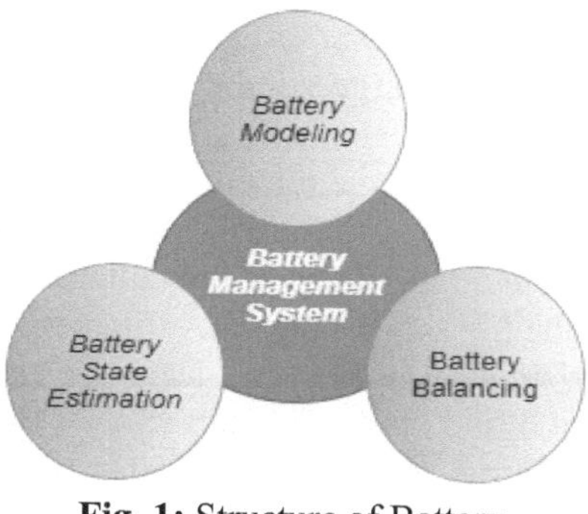

Fig. 1: Structure of Battery Management System

BMS describes experimental methods and model based methods for the estimation of actual capacity of the battery. In experimental methods, available capacity of the battery can be calculated at the battery terminals [1]. Voltage at battery terminals can be evaluated mathematically when battery is charged at constant current or constant voltage that is amenable to estimating capacity of it. Experimental methods calculate capacity of the battery from open circuit voltage method and coulomb counting method. However, model based methods need any mathematical model to estimate the capacity of the battery which depend on the thought of non-linearity. Mathematical models are much efficient to give an accurate capacity estimation of the battery.

Battery Modeling

Battery modeling basically consists of a mathematical model which forms an electrical circuit having passive elements such as resistor, inductor and capacitor connected with a battery. A battery model is designed to operate the vehicle and evaluate the state of the battery. A well-designed battery model enhances the efficiency and the life cycle of the battery [2].

A battery model consists of an electrochemical system in which the both electrodes anode and cathode are immersed in an electrolyte. The very first approach for an electrochemical system is a black-box approach that uses current and voltage characteristics to design a mathematical model.

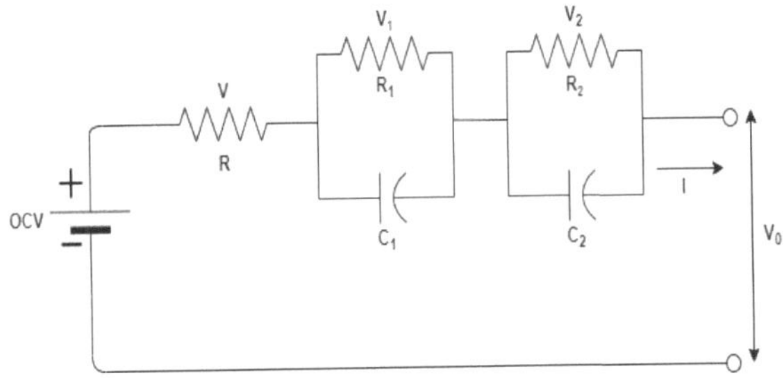

Fig. 2: Equivalent Circuit Model for a Battery

The second approach for an electro-chemical system is a lumped-model approach for rapid measurements. Further, thebattery model introduces to the reduced-order battery model, equivalent circuit battery model and data – driven battery model.

An equivalent circuit battery model contains a circuit in which the voltage source is connected to the resistors and capacitors. A simple equivalent circuit model for a battery is shown in the above Figure 2. The equivalent circuit consists of a RC model which was presented by the National Renewable Energy Laboratory (NREL) [3]. A parallel RC network is utilized to resolve the dynamic response of the battery. In RC model, R and C expresses the concentrated and electro-chemical polarization respectively.

Here, OCV = Open circuit voltage.
V = Voltage across the internal resistance.
V_1 = Voltage across the parallel R_1C_1 circuit
V_2 = Voltage across the parallel R_2C_2 circuit
V_0 = Battery terminal voltage.

The mathematical equation for an equivalent circuit of a battery is given as:

The internal resistance is,

$$R = \frac{V-OCV}{I}$$

Where, I is the discharging current and $V_1 - OCV$ is the voltage difference.

The mathematical equations for identifying the R_1, R_2, C_1 and C_2 from the discharging current (I). That is given as:

$$C1\frac{dV1}{dt} + \frac{V1}{R1} = I$$

$$C2\frac{dV2}{dt} + \frac{V2}{R2} = I$$

All of the above equations express the equivalent circuit battery model.

Battery State Estimation

State of a battery can be defined as a minimum amount of information that drive the charge of the battery from initial value at time (t_0) to the final value at a finite time (t_f). A battery management system is engaged to determine the states of the battery used in the EVs. A battery can estimates two different forms of state—state of charge (SoC) and state of health (SoH). Both the terms are helpful for the battery calculation and enhancing the cycle life of the battery [4].

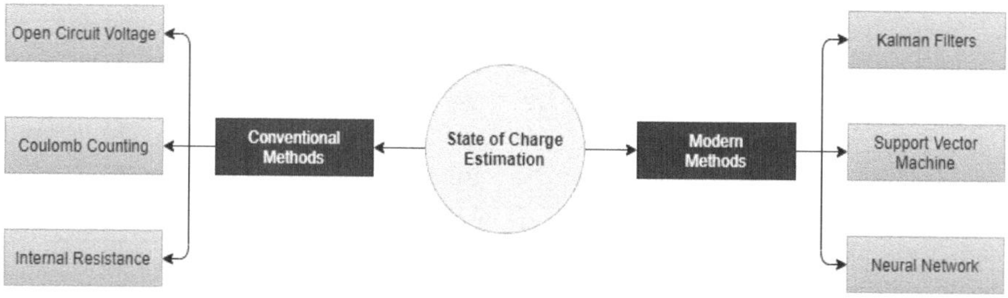

Fig. 3: Different Methods for SoC Estimation

A. *State of Charge Estimation*

State-of-Charge (SoC) introduces to the available capacity of the battery. It can be defined as the ratio of the current capacity to the maximum or nominal capacity of the battery. It is expressed in terms of percentage. If the SoC is 100% that means the battery is fully charged and if the SoC is 0% that means the battery have no available capacity to supplying electricity to the vehicle [2]. The formula for the state-of-charge of a battery can be given as:

$$SoC = \frac{Q(t)}{Qn}$$

Where, $Q(t)$ is the current capacity and Q_n is the total capacity of a battery.

- ***Open Circuit Voltage Method:*** Open circuit voltage (OCV) is a direct method for SoC estimation which shows a linear-characteristics between state of charge and its open circuit voltage [3]. In this method, both the voltage and the SoC simultaneously observed to draw a small discharge current from the battery. A polynomial equation is obtained from the relationship between OCV and the SoC,

$$u\,(OCV) = f\,(SoC)$$

A lithium-ion battery does not show linear characteristics between state of charge and its open circuit voltage of a battery.

- ***Coulomb Counting Method:*** It is current based state of charge estimation method. Coulomb counting is the broadly used SoC estimation method that analyses the discharging current from the battery. This method provides a little error in the evaluation of discharging current [5]. It is also called the current integration method. The estimated SoC by using coulomb counting method can be given as,

$$SoC = SoC(t-1) + \frac{i(t)}{Qn}\Delta t$$

Where, SoC $(t\text{–}1)$ is the previous estimated SoC value and $i(t)$ is the discharging current.

- ***Internal Resistance Method:*** The SoC can be estimated from the relationship between the internal resistance of the battery and SoC value. The SoC relies the relationship between the internal resistance and voltage of the battery. Such a relationship comes from the open circuit voltage. This method is not applicable for the state of charge estimation of a lithium-ion battery pack [5].

- ***Kalman Filters Method:*** It is a model-based method of estimating SoC among to the various state of charge estimation methods. The Kalman Filters method attracts a huge attention to evaluate the SoC from the equivalent circuit model with the increment of the complexity in the parameter's computation. The equivalent circuit model provides equations for the state [6].

  The Kalman Filters method describes the error between actual voltage of the battery and the voltage from the equivalent circuit. Hence, this method establishes the equation for the minimum variation in the SoC estimation. The Kalman Filters method is able to sift the harmonics and several kinds of errors to access the state estimation.

Above methods are conventional methods for estimating the state of charge. Now-a-days, Artificial Intelligence is popularly used technique that gives an accurate estimation of the SoC and the charging-discharging behaviour of the BMS. Artificial Intelligence (AI) uses artificial neural network algorithm and fuzzy logic topology to estimate the state of charge [6].

Moreover, the state of charge introduces to another term that is depth of discharge (DoD). The depth of discharge is defined as the inverse of the state of charge. It expresses the capacity that is being discharged from the battery. The term DoD represents the life cycle of a battery. The formula for depth of discharge is given as:

$$DoD = \text{Capacity} \times \frac{100 - SoC}{100}$$

The DoD is also an important term that can be calculated from the estimation of the state of charge.

B. *State of Health Estimation*

State-of-Health (SoH) can be defined as the ratio of the available capacity of a battery in currently situation to the maximum or nominal capacity of a new and fresh battery. It is expressed in terms of percentage. SoH of the battery can be estimated through the degraded capacity or increased internal resistance [7]. The formula for the state-of-health of the battery can be given as:

$$SoH = \frac{C(\text{degraded})}{Cn} \text{ or } \frac{R(\text{increased})}{Rn}$$

Where, the degraded capacity and the increased internal resistance of the currently used battery. C_n is the nominal capacity and R_n is the internal resistance value of the fresh battery.

State-of-Health (SoH) can be estimated by using different methods similar to the SoC estimation methods. A schematic representation of the different methods of SoH estimation is shown in the above Figure 4.

State-of-Health can be evaluated from the direct methods such as open circuit voltage, coulomb counting, internal resistance methods etc. These methods provide SoH estimation from

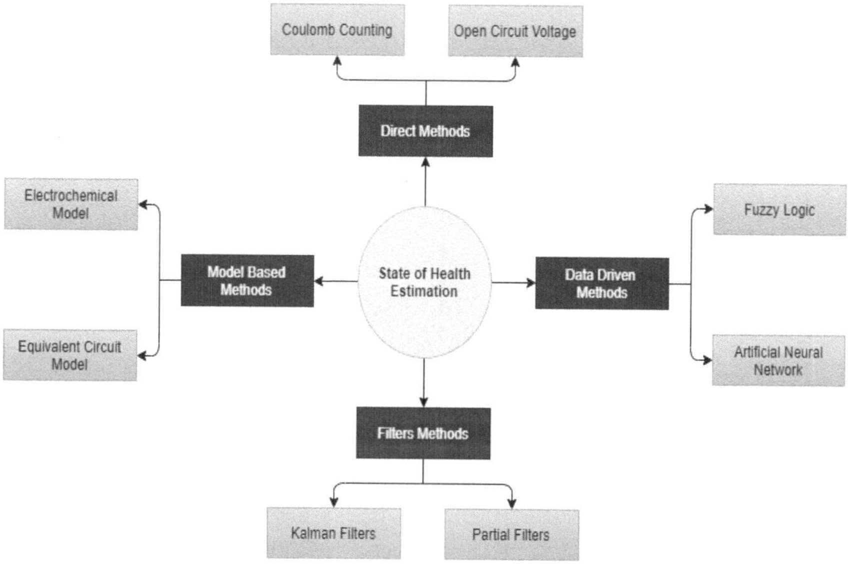

Fig. 4: Different Methods for SoH Estimation

the monitoring of the battery terminal voltage, charging-discharging current and the internal resistance of the battery [4]. Model-based methods for SoH estimation introduces to the electrochemical methods and equivalent circuit model. The electrochemical method relies on the chemical reaction between the electrodes and an electrolyte used in the battery. Whereas the equivalent circuit model consists of resistors, capacitors and voltage source that help for fast computation for the state of health [7].

The estimation methods for the state of health are similar to the SoC estimation methods. In the filters methods of SoH estimation, the Kalman filters and partial or particle filters are used for accurate evaluation of SoH. Further, data driven methods are also elaborated that can be using fuzzy logic, artificial neural network, support vector machine etc. These methods are developed computation methods that provide the SoH estimation in graphical form. These computation methods use algorithm and complex models for the accurate estimation of the state-of-health [7].

A relationship is established between the state of charge and the state of health in terms of the depth of discharge. It is given as:

$$SoC\ (t) = SoH\ (t) - DoD\ (t)$$

It is the state estimation for the battery engaged in the electric vehicles.

Battery Balancing

A battery pack having a large number of electric cells connected in series to enhance the total voltage supplied from the battery. However, the battery

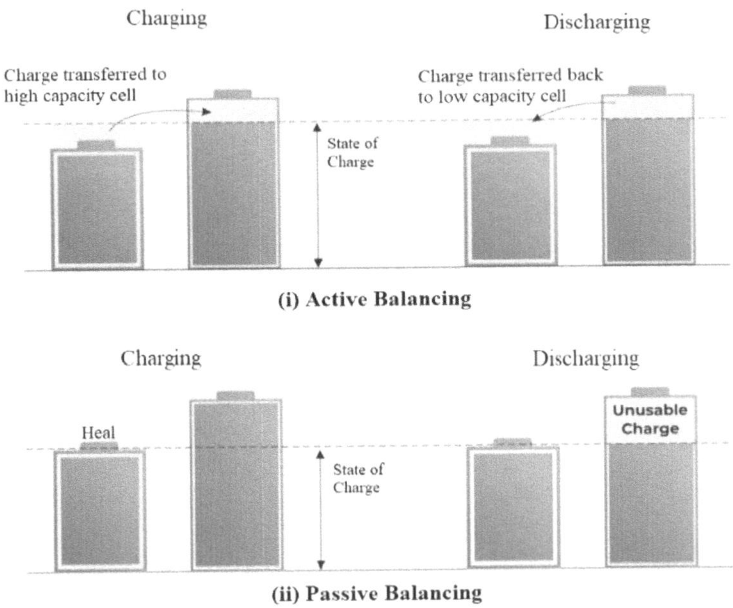

Fig. 5: Cell Balancing Methods—Active and Passive Balancing

parameters such as internal resistance, cycle life, capacity etc. are not same for all the electric cells. Due to the imbalance in all the parameters, some of the weak cells begin to be overheated and overstressed during the charging period. Sometimes, the battery over-charges or over discharges due to the imbalance in the parameters [8]. Therefore, there is a need to balance all the parameters of the battery cells.

Battery balancing contains significant importance to maintain and equalize all the battery parameters of the cells in the battery. It is also known as cell balancing which improves the life cycle and the performance of the battery [9]. It maintains the total energy among all the cells used in a battery pack. Further, the battery balancing introduces to the active and passive balancing methods as shown in the Figure 5 above.

In active balancing method, the energy can be transferred between the electric cells by using the switching circuits. This method has wasted energy in a little amount. So that, the number of switches is increased to increase the excessive energy. The active balancing method is costly and unreliable. Whereas in passive balancing method, a switching resistor is used for energy dissipation in from of heat. The excessive energy is also collected by using a switching resistor. The passive balancing methods raise the complexity in temperature control. The passive balancing method is easy to understand, low in cost and reliable [10].

Artificial Neural Network Strategy for State-of-Charge Estimation

A lithium-ion battery is used for performing the estimation of available capacity throughout the ANN strategy in MATLAB [11]. The specification for the lithium-ion battery is shown in a Table 1 as:

Table 1: Lithium-ion Battery Specifications

Type of Battery	*Lithium-ion Battery*
Nominal Voltage	72 V
Rated Capacity	20 Ah
Fully Charged Voltage	83.80 V
Cut-off Voltage	54 V
Internal Resistance	0.036 Ω

Artificial neural networks (ANNs) have ability to perform computation with flexibility and power of human brain. ANN acts as a controller which observes and monitors the working operations of a system. Basic signal processing element of a neural network is known as neuron. In a neural network, thousands of neurons are connected to each other and each connection is shown by a numeric digit that is known as 'weight'. The neural networks are implemented as a non-linear control for the proposed lithium-ion battery management system [12]. Input layer represents linguistic variables and the hidden layer shows membership function of the input linguistic variables. Hence, an output is achieved in output layer from the applied input variables [13].

A basic architecture of an artificial neural network is shown in the Fig. 6.

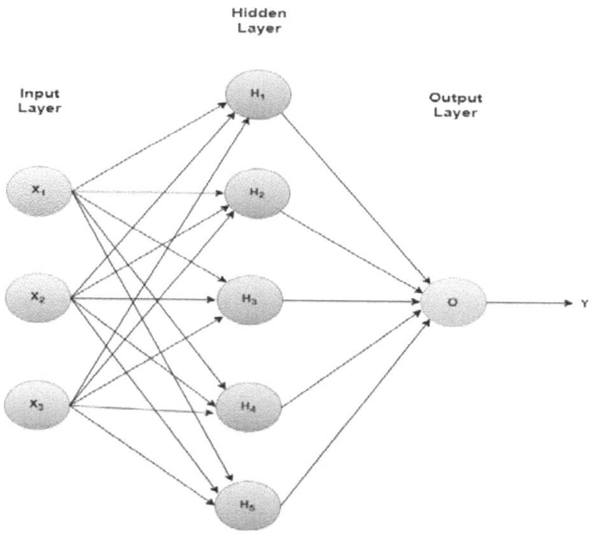

Fig. 6: Architecture of Artificial Neural Network

Artificial neural network approach for the determination of state of charge for a lithium-ion battery management system. Modeling of ANN can be performed by using NN tool available in MATLAB. The input layer for the neural network contains three inputs for a lithium-ion battery for example battery voltage, battery current and battery temperature [14]. The available capacity of the proposed battery can be estimated in the output layer from the given inputs.

Simulation Response of the Neural Network

State of charge can be evaluated from the given inputs i.e. voltage, current and temperature by using artificial neural network strategy. Initially, the neural network is trained from the 70% of all of the input and target data. Further, the neural network will be tested by applying the 30% of all of the input and target data [15]. The neural network is designed, trained and tested at two temperature –10°C and 25°C. The training results of the designed network comprises of three parts—performance, training state and regression at 100 epochs.

Performance of the network shows the value of the performance function versus number of epochs. The best validation performance is 0.0023215 at 0 epoch. The Figure 7 describes the straight and minimal mean squared error between all the three lines of the network as (Figure 7).

Training state curve in Figure 8 shows the progress of other training variables such as gradient magnitude, training gain and the number of validation checks. Through the results of training state, the gradient magnitude is 3.3631×10^{-5}, training gain is 1×10^{-7}, validation checks are counted 100 at epoch 100. All the three training variables at 100 epochs are shown as (Figure 8).

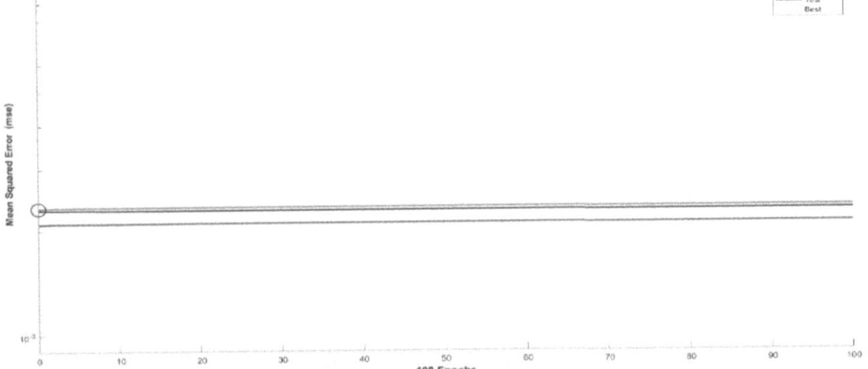

Fig. 7: Performance of the Network

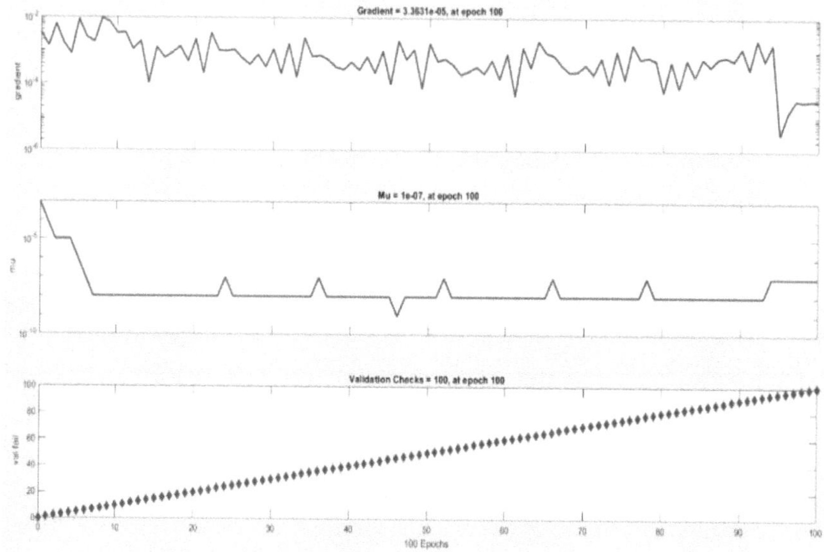

Fig. 8: Training State of the Network

Regression of the neural network shows a regression graph between network's outputs and network's targets. For regression task, the regression value can be one value. Linear regression shows a line of best fit inside the datasets. The regression results comprised of four parts—training, validation, testing and all dataset. All the four results of the regression of the network are shown in the Figure 9.

After the training of the neural network, 30% of the all the input and target data will be considered as testing data. After the testing of the neural network, available capacity is estimated for a lithium-ion battery-based management system. State of charge at –10°C is estimated with more accuracy. Hence, the artificial neural network is helpful for estimation of available capacity in the li-ion battery with more accuracy.

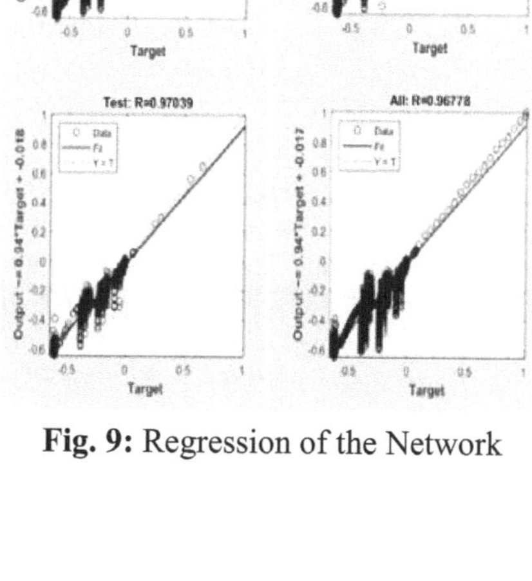

Fig. 9: Regression of the Network

Conclusion

In the paper, simulation of a lithium-ion battery management system has been discussed. MATLAB simulation is used to perform battery management system design and simulation response is calibrated for determination of available capacity through artificial neural network approach. ANN strategy is motivated on BMS of lithium-ion battery in which the best validation performance is measured about 0.0023215 at 0 epoch.

By performing training and testing of the neural network created using NN tool available in MATLAB, we can conclude that the available capacity is determined with more accuracy from the given inputs i.e. voltage, current and temperature. State of charge (SoC) at temperature of –10°C is estimated accurately and rapidly as compared with the State of charge (SoC) at temperature of 25°C.

References

[1] Xu, J. and Cao, B. (2015). Battery management system for electric drive vehicles: Modelling, state estimation and balancing. *New Applications of Electric Drives, 1st ed.; Chomat, M., Ed*, pp. 87–113.

[2] Brandl, M., Gall, H., Wenger, M., Lorentz, V., Giegerich, M., Baronti, F. and Saponara, S. March (2012). Batteries and battery management systems for electric vehicles. *Design, Automation & Test in Europe Conference & Exhibition IEEE*. pp. 971–976.

[3] Shen, W.X., Chan, C.C., Lo, E.W.C. and Chau, K.T. (2002). A new battery available capacity indicator for electric vehicles using neural network. *Energy Conversion and Management*, 43(6): 817–826.

[4] Gazzarri, J., Shrivastava, N., Jackey, R. and Borghesani, C. (2014). Battery pack modeling, simulation, and deployment on a multicore real time target. *SAE International Journal of Aerospace*, *7*(2014-01-2217): 207–213.

[5] Ren, H., Zhao, Y., Chen, S. and Wang, T. (2019). Design and implementation of a battery management system with active charge balance based on the SOC and SOH online estimation. *Energy*, 166: 908–917.

[6] Liu, K., Li, K., Peng, Q. and Zhang, C. (2019). A brief review on key technologies in the battery management system of electric vehicles. *Frontiers of mechanical engineering,* 14(1): 47–64.

[7] Vidal, C., Kollmeyer, P., Naguib, M., Malysz, P., Gross, O. and Emadi, A. (2020). Robust xev battery state-of-charge estimator design using a feed-forward deep neural network. *SAE International Journal of Advances and Current Practices in Mobility*, 2(2020-01-1181): 2872–2880.

[8] Hussein, A.A. (2014). Capacity fade estimation in electric vehicle li-ion batteries using artificial neural networks. *IEEE Transactions on Industry Applications*, 51(3): 2321–2330.

[9] Xie, S., Hu, X., Qi, S. and Lang, K. (2018). An artificial neural network-enhanced energy management strategy for plug-in hybrid electric vehicles. *Energy,* 163: 837–848.

[10] Li, S. and Zhang, C. (2009). March. Study on battery management system and lithium-ion battery. In *2009 International Conference on Computer and Automation Engineering IEEE,* pp. 218–222.

[11] Man, K.L., Chen, C., Ting, T.O., Krilavičius, T., Chang, J. and Poon, S.H. (2012). Artificial intelligence approach to SoC estimation for smart BMS. *In: ECT-2012 Electrical and control technologies: proceedings of the 7th international conference on electrical and control technologies, Kaunas, Lithuania during* May 3–4, 2012.

[12] Omariba, Z.B., Zhang, L. and Sun, D. (2018). Review on health management system for lithium-ion batteries of electric vehicles. *Electronics*, 7(5): 72.

[13] Rimon, K.S.M.L., Sulaiman, S.I. and Omar, A.M. (2019). System protection for lithium-ion batteries management system: A review. *Indonesian Journal of Electrical Engineering and Computer Science (IJEECS)*, 13.

[14] Sen, C. and Kar, N.C. (2009). Battery pack modeling for the analysis of battery management system of a hybrid electric vehicle. *IEEE Vehicle Power and Propulsion Conference*, pp. 207–212.

[15] Zhang, F. (2017). Modeling and control of a modular battery management system for lithium-ion battery packs. Doctoral dissertation submitted to University of Colorado, Boulder, United States.

Study of Energy Management for Hybrid Energy Storage System Connected to Standalone PV based Microgrid

Ayushi[1] and Ram Niwash Mahia[2]
Department of Electrical Engineering,
National Institute of Technology Hamirpur,
Hamirpur–177005, India
E-mail: [1]20mee004@nith.ac.in; [2]ram@nith.ac.in

ABSTRACT: *In this paper, a novel filtration based Energy Management Strategy (EMS) is analyzed for Hybrid Energy Storage System (HESS) connected to PV system. Due to its low power density battery can't handle the sudden increase in load demand but supercapacitor (SC) can. Because of it's high power density. So there is a need for efficient energy management strategy which can distribute the slow changing power surges and fast changing power surges in the two energy storage units. Because battery can only handle the slow changing power surges and supercapacitor is good at handling fast changing power surges. In standalone PV generation based microgrid an energy buffer is required which can handle the power mismatch between total generation and required demand. A method based on low pass filter for energy management between the two storage units is analyzed and the results are discussed in this paper.*

Keywords: EMS (Energy Management strategy); HESS (Hybrid Energy Storage System); PV (Photovoltaic); Battery; Supercapacitor (SC).

Introduction

Renewable energy storage system is considered as best option for remote area power supply. Because of their lesser cost and easy installment in comparison to the conventional resources, Photovoltaic and wind generation is considered best among all other renewable energy resources because of their simplicity and less cost [3].

Photovoltaic systems have intermitted power generation because of the change in some factors like weather, season and time. The power generation changes with change in temperature and irradiance due to which it is not efficient to supply load demand. A way of maintaining the power supply to the load is through an energy storage system that can provide additional power. And the surplus energy from photovoltaic source can be supplied to the energy storage system and can be used when needed [4].

Battery is having high energy density, lower power density and slow charge or discharge rate. Due to the variable power generation of the photovoltaic system battery deals with irregular charging and discharging pattern sometimes battery gets under-charged which leads to sulfation and stratification and sometimes it gets overcharged which leads to gassing because of these factors the battery life gets shortens [1].

High power density and low energy density of supercapacitors make them the most suitable for peak power demand; an energy storage system should have both qualities. The super-capacitor is having high power density but low energy density and higher charge/discharge rate. This makes them best suitable for supplying peak power to the load. A good energy storage unit should have both qualities high energy density and high power density. If alone battery is used then we have to use an oversized battery for supplying the peak load. In order to store energy in renewable energy resources, HESS with battery and supercapacitor provides the best solution because of their complementary characteristics: the combination of both has high energy density and high power density. But battery can only handle the slow changing power surges it can't handle the fast changing power surges and super-capacitor is good at handling fast changing power surges. So there is a need of good energy management strategy which can divert the fast changing power surges from battery to the supercapacitor. By using an efficient EMS several advantages can be achieved like reduction in battery size and cost, reduction in battery burden and increment in lifetime of battery, improvement in power balance between power generation and load demand [2]. The DC bus voltage changes from its reference value with change in generation and load. In the event that the load demand exceeds the power generation, then the DC bus voltage will decrease from its reference value. Conversely, in the event that the power generation exceeds the load demand, the DC bus voltage will increase from its reference value [2].

There are many advantages of using a HESS instead of a single battery like power quality improvement if connected to AC grid, DC bus voltage regulation in DC microgrid, With HESS, the power generated balances with the load demand, reducing current stresses on battery, increases the life cycle of battery, we don't have to frequently replace battery, there will be less burden on battery [10].

A brief discussion about the modeling of HESS connected to PV based microgrid which includes the study of all the step is presented [4], [10], [11], [19]. Different strategies have been reported in [3], [13] [18] [17]. A Decentralized Dynamic Power

Table 1: Battery versus Supercapacitor [2]

	Lead Acid Battery	*Supercapacitor*
Specific energy density	10–100 Wh/kg	1–10 Wh/kg
Specific power density	< 1000 (W/kg)	< 10,000 (W/kg)
Cycle life	1000	> 500000
Efficiency	70–85%	85–98%
Fast charge time	1–5 hr	0.3–3 sec.
Discharge time	0.3–3 hr	0.3–3 sec.

Sharing Strategy for Hybrid Energy Storage System in Autonomous DC Microgrid is developed in [20]. Control strategies based on the low pass filter and SOC (state of charge) regulation is proposed [5].

This paper studies a novel energy management strategy for the energy storage system of PV Array which is connected to a dc bus. And results are presented and discussed here.

This paper is organized as follows. In section II system configuration is briefly discussed which includes the modeling of the components used. In section III low pass filter based energy management strategy is described and the simulink model of the system is shown. In section IV the results and the simulation is presented and analyzed. Section V followed by conclusion.

In Table 1 comparison between the characteristics of battery and super-capacitor is presented.

System Configuration

Figure 1 shows the block diagram of system configuration where PV unit is connected to the DC bus through boost converter and the duty cycle to the converter is provided by PWM generator using MPPT algorithm. The boost converter is used to extract maximum power from the PV array. The battery and supercapacitor are connected to the DC bus using bidirectional DC-DC converter for bidirectional power flow between dc bus and energy storage elements. So that when the generation will be more than demand power extra power can be supplied to ESS (energy storage system). If the demand is more than generation then power can flow from ESS to the load. The switching signal to battery SB1, SB2 (for battery) and the switching signal to supercapacitor SS1,

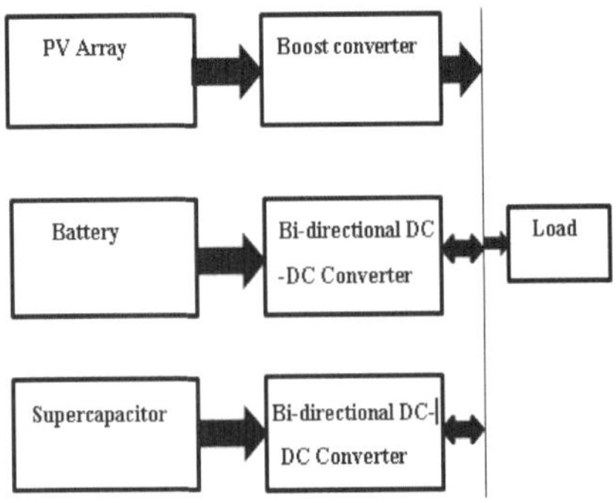

Fig. 1: System Configuration

SS2 (for supercapacitor) are supplied through the output of energy management system output. The total load connected to the system as shown in Figure 1 is taken as an equivalent resistance *R*.

PV Array

PV cells are the fundamental building block of PV systems which laminated in environmentally protective layer. PV cells are connected in series or parallel to provide high current or voltage. The PV module basically converts the solar power into electrical energy by photovoltaic effect. The characteristics equation of PV array can be given as [6], [13].

$$I_{pv} = N_p I_{ph} - N_p I_{rs}\left(exp\left(\frac{q\left(\frac{V_{pv}}{N_S}+\frac{I_{pv}R_S}{N_p}\right)}{AkT}\right)-1\right) - \frac{N_p\frac{V_{pv}}{N_S}+I_{pv}R_S}{R_{sh}} \quad \ldots (1)$$

Equation 1 for PV array have terminal current (I_{pv}), photocurrent (I_{ph}), reverse saturation current (I_{rs}), Terminal voltage (V_{pv}), electron charge (q), Boltzmann's constant (k) and absolute cell temperature (K), Np is number of cell in parallel and Ns is number of cell connected in series. Refer to [7].

Battery

Battery can be modelled using the following equation [8].

$$E = E_o - K\frac{Q}{(Q-\int idt)} + Ae^{-B\int idt} \quad \ldots (2)$$

$$V_{batt} = E - Ri \quad \ldots (3)$$

$$\%SOC = \left(1 - \frac{1}{Q}\int idt\right) \times 100 \ldots (4)$$

Where I = terminal current of battery and V_{batt} = terminal voltage of battery [16].

Supercapacitor

Supercapacitor is also known as ultracapacitor or double layer capacitor. They are having capacitances in the range of thousands of farads. In a supercapacitor, the electrodes are immersed in an electrolyte and a separator that allows ions from the electrolyte to pass through but prevents electrical contact is implanted between the electrodes.

Supercapacitors are having much higher lifecycle than battery. It store energy like normal capacitor by separating charges into positive ion and negative ions, but battery stores charge chemically. It also have high charge/discharge rate and can store at any value of charge. Figure 2 shows the equivalent model of SC [9], [15].

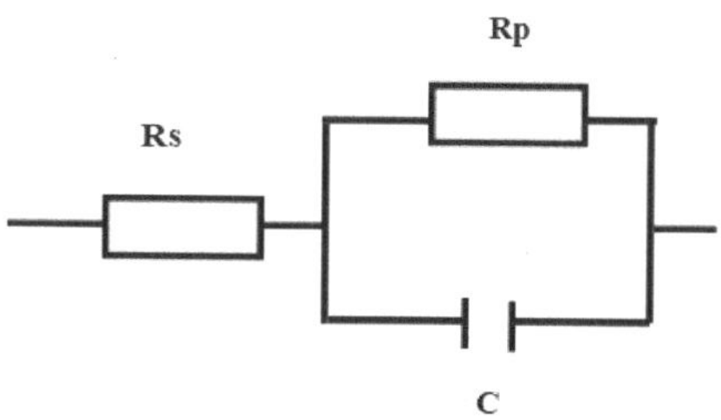

Fig. 2: Equivalent Representation of Supercapacitor

Battery-Supercapacitor Converter Topology

Battery and supercapacitor are connected to the bus through bidirectional dc-dc converter so that power can flow in both directions whichever is required. Following are the types of topologies which are used for connecting energy storage elements to the DC bus.

(a) Passive topology
(b) Semi active topology
(c) Active topology

(a) *Passive Topology*

In passive topology battery-supercapacitor are connected in series then directly connected to the DC bus and have same terminal voltage as shown in Figure 3, in this configuration SC behaves like a low pass filter, Advantage of using this topology is that it is very easy to implement and less costly than other topologies and disadvantage is that the distribution of power between SC and battery is not controlled its totally depends on the internal their internal resistances [3].

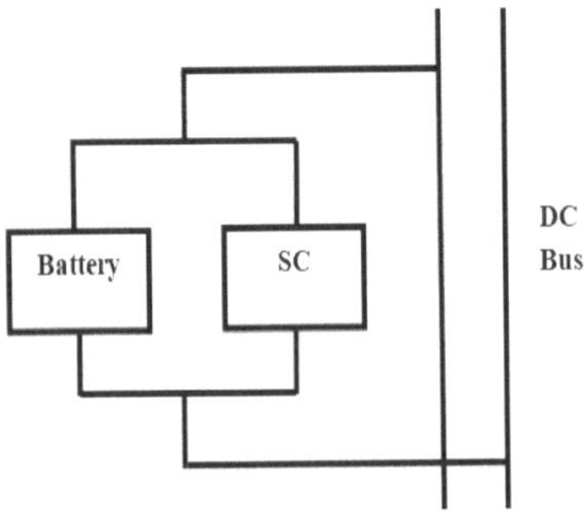

Fig. 3: Passive Topology

(b) *Semi Active Topology*

In semi active topology battery and supercapacitor are connected in parallel to the DC bus. Only one converter is connected to the energy storage element, either to battery or to the supercapacitor as shown in Figure 4. In semi active topology only one of the ESS unit can be actively controlled [10].

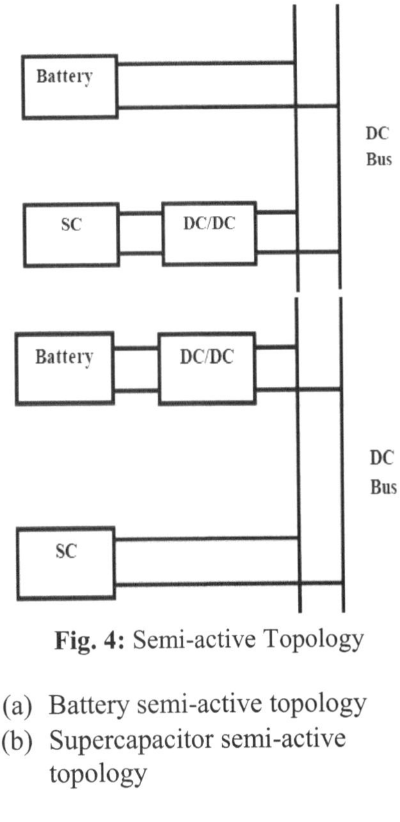

Fig. 4: Semi-active Topology

(a) Battery semi-active topology
(b) Supercapacitor semi-active topology

(c) *Active Topology*

In active topology two converters are used, battery and supercapacitor each one is connected to single-single converter and connected to DC bus either in series or in parallel connection as shown in Figure 5. This topology is costliest because two converters are used here but both of the energy storage units can be actively controlled.

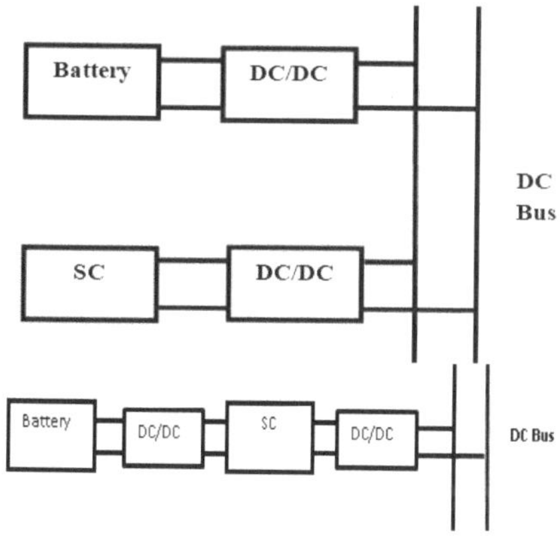

Fig. 5: Active Topology
(a) Parallel active topology
(b) Cascaded active topology

Energy Management Strategy

Figure 6 shows the energy management strategy which is analyzed here. Basically the reference current divides into two parts, low frequency term and high frequency term which is done by low pass filter. Basically the low pass filter does it. The DC bus voltage is compared to reference DC bus voltage and error signal is supplied to the PI Controller (Proportional Integral controller) which gives the total reference current required from the HESS.

The output of the PI controller passed through the low pass filter. The low frequency component of the total reference current required from HESS compared to the reference current of battery. Obtained error signal is now passed through PI controller. Then supplied to PWM generator. PWM generator gives the switching pulses SB1 and SB2 to battery. The high frequency component of the total reference current can be given as,

$$I_{HFC} = I_{ref} - I_{LFC} \quad \dots (5)$$

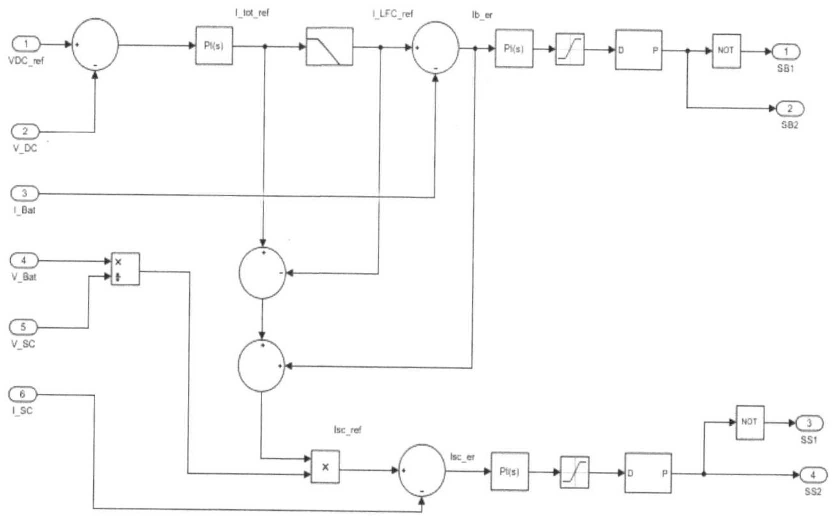

Fig. 6: EMS (Energy Management Stratergy)

Due to the slow response of battery the battery can't compensate the total power so there will be some un-compensated power which is given as,

$$P_{\text{uncompensated}} = (I_{HFC} + I_{B_{err}}) \times V_B \quad \dots (6)$$

This uncompensated power to be compensated by SC the SC reference current can be given as,

$$I_{S\_ref} = \frac{P_{B\_\text{uncompensated}}}{V_S} = (I_{HFC\_ref} + I_{B\_err}) \times \frac{V_B}{V_S} \quad \dots (7)$$

The uncompensated power is compensated by supercapacitor. Then reference supercapacitor current is compared to supercapacitor feedback current and the error signal is provided to the PI controller, then it is passed through the PWM generator, from where switching signal is generated for battery and supercapacitor. Outputs signals SB1, SB2, SS1 and SS2 are provided as switching signal to battery and supercapacitor. SB1 and SB2 are for battery and signals SS1 and SS2 are for supercapacitor.

Simulation and Results

A Simulink model which is shown in Figure 6 is simulated with the help of Simulink software. The used system specifications are given in Table 2. After simulating this model in simulink graph observed and the obtained results are discussed in this section.

Table 2: Specifications

Subsystem	*Specifications*
PV Array	V_{pv} = 40V, I_{pv} = 30A
Battery	Lithium Ion, Rated Voltage = 24 V, Rated Current = 14 Ah
Super-capacitor	Rated capacitance = 29 F, Rated voltage = 32 V
DC-DC Converters	L_{pv} = 0.352 mH, L_B = 0.3 mH, L_S = 0.355 mH, C = 2200 μC, R = 0.05

Change in temperature and irriadiance values with time are shown in Figure 7 where temperature is constant at 25 degree celcius and initial

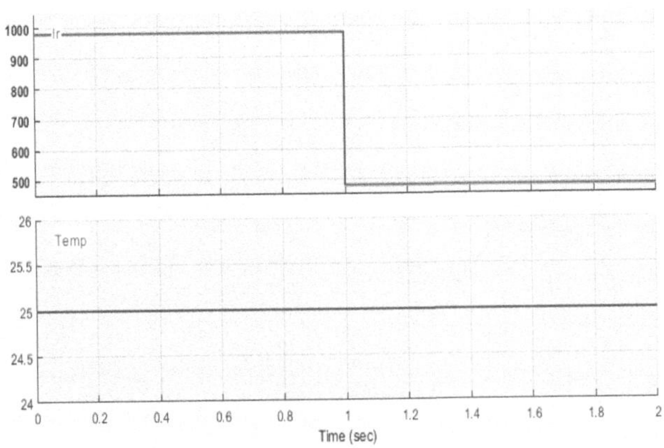

Fig. 7: Solar Irradiance and Temperature

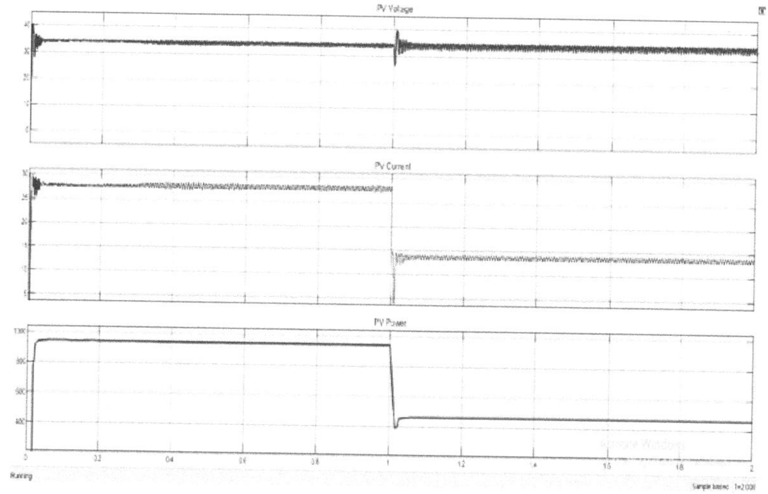

Fig. 8: PV Array Specifications—PV Voltage, Current and Power

value of irradiance is 1000 W/m^2 after 1 second irriadiance is changed to 500 W/m^2. PV generated power also gets changed with change in irradiance because of the intermitted nature of it.

In Figure 8 effect of change in irradiance in PV current, voltage and power is plotted. PV current and power gets reduced from its initial value. Due to this reduction in generation DC bus voltage also reduces from its reference value and load demand also not fulfilled.

Plot of different powers are shown in Figure 9 where at t = 1 sec. solar irradiance changes and generation gets reduced. Some extra power generation is needed so that required power could

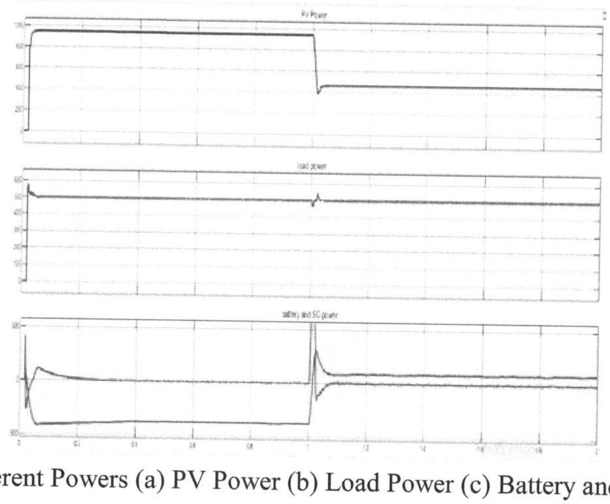

Fig. 9: Different Powers (a) PV Power (b) Load Power (c) Battery and SC Power

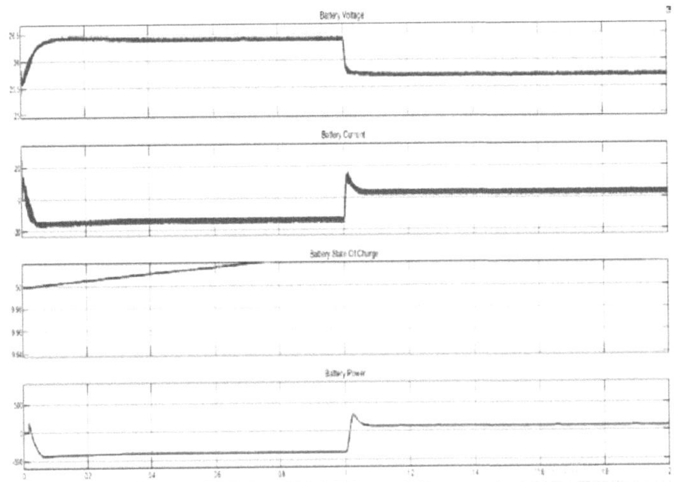

Fig. 10: Battery Voltage, Current, State of Charge and Power

be supplied to the load. This sudden decrease in generation is compensated by energy storage system. Super-capacitor provide power at the same instant and battery for a long time.

Effect of change in irradiance in battery parameters can be understood by Figure 10. The instant power generation decreases power supplied by battery increases and this increment is for a long duration. Because battery is capable of supplying power for longer duration in comparison to SC. And the changes in SC parameters can be understood by Figure 11 when the power generation reduces SC supplies

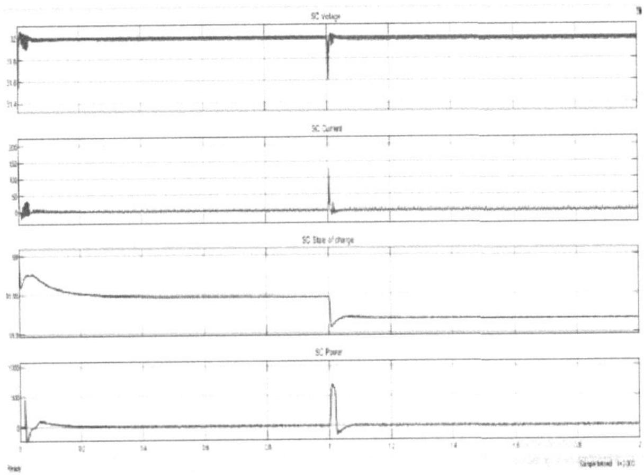

Fig. 11: Supercapacitor Voltage, Current, State of Charge and Power

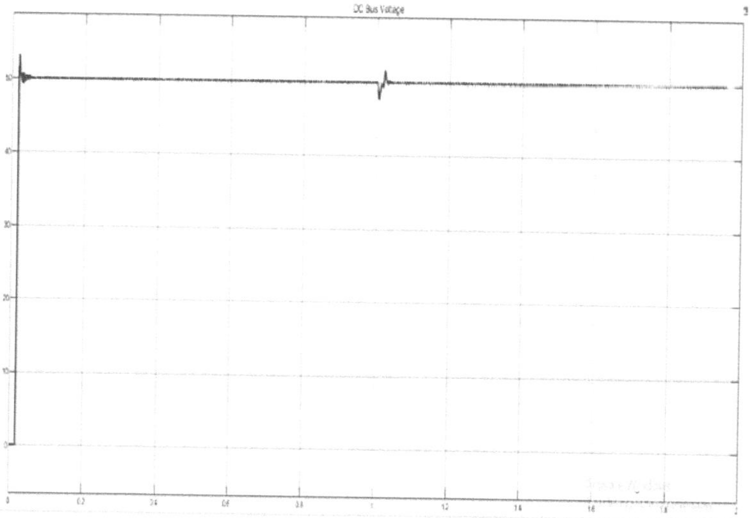

Fig. 12: DC Bus Voltage

the additional power with battery from the ealier stored power.

Due to change in irradiance DC bus volatge also fluctuates from its reference value but because of the application of proper energy management stratergy it is maintaied at the reference value which is set at 50 V. irrespective of change in generated power from PV array. Plot of DC bus voltage with respect to time is shown in Figure 12.

Conclusion

Analysis of a filter based energy management strategy is done in this paper. This strategy is basically based on monitoring of generation-demand mismatch and divides into low frequency and high frequency terms. This method is very simple for implementation and need less computation. This method uses the error signal of battery for controlling supercapacitor so that the high frequency terms could be directed to supercapacitor and low frequency to battery. High frequency terms are handled by SC and low frequency terms are handled by battery. This method successfully manages power between battery and SC. In this paper several advantages can be obtained using this method like faster DC bus voltage regulation, lesser current stresses on battery, reduced size of battery, lower charge/discharge rates of battery, increased life cycle of battery, less computational burden.

References

[1] Glavin, M.E. and Hurley, W.G., "Ultracapacitor/battery hybrid for solar energy storage." *2007 42nd International Universities Power Engineering Conference*. IEEE, 2007.

[2] Kollimalla; Kumar, Sathish; Mishra, Mahesh Kumar and Narasamma, N.

Lakshmi, "Design and analysis of novel control strategy for battery and supercapacitor storage system." *IEEE Transactions on Sustainable Energy* 5.4 (2014): 1137–1144.

[3] Cabrane, Zineb, *et al.* "HESS-based photovoltaic/batteries/supercapacitor s: Energy management strategy and DC bus voltage stabilization." *Solar Energy* 216 (2021): 551–563.

[4] Hajiaghasi, Salman, Ahmad Salemnia and Hamzeh, Mohsen, "Hybrid energy storage system for microgrids applications: A review." *Journal of Energy Storage* 21 (2019): 543–570.

[5] Kollimalla, Sathish Kumar and Mishra, Mahesh Kumar. "Co-ordinated control and energy management of hybrid energy storage system in PV system." *2014 International Conference on Computation of Power, Energy, Information and Communication (ICCPEIC). IEEE*, 2014.

[6] Joyce, A., Rodrigues, C. and Manso, R. "Modelling a PV system." *Renewable energy* 22.1–3 (2001): 275–280.

[7] González-Longatt, Francisco M. "Model of photovoltaic module in Matlab." *Ii Cibelec* 2005 (2005): 1–5.

[8] Tremblay, Olivier, Louis-A. Dessaint and Abdel-Illah Dekkiche. "A generic battery model for the dynamic simulation of hybrid electric vehicles." *2007 IEEE Vehicle Power and Propulsion Conference*. IEEE, 2007.

[9] Fărcaş, C., *et al.* "Modeling and simulation of supercapacitors." *2009 15th International Symposium for Design and Technology of Electronics Packages (SIITME)*. IEEE, 2009.

[10] Jing, Wenlong, *et al.* "Battery-supercapacitor hybrid energy storage system in standalone DC microgrids: areview." *IET Renewable Power Generation,* 11.4 (2017): 461–469.

[11] Hajiaghasi, Salman; Salemnia, Ahmad and Hamzeh, Mohsen, "Hybrid energy storage system for microgrids applications: A review." *Journal of Energy Storage,* 21(2019): 543–570.

[12] Kumar, Jaynendra, Anshul Agarwal and Agarwal, Vineeta. "A review on overall control of DC microgrids." *Journal of energy storage,* 21 (2019): 113–138.

[13] Tsai, Huan-Liang; Tu, Ci-Siang and Su, Yi-Jie, "Development of generalized photovoltaic model using Matlab/Simulink." *Proceedings of the world congress on Engineering and computer science,* Vol. 2008, 2008.

[14] Javed, Kashif, *et al.* "Design and performance analysis of a stand-alone PV system with hybrid energy storage for rural India." *Electronics,* 8.9 (2019): 952.

[15] Perdana, Yoga Sandi, *et al.* "Direct connection of supercapacitor–battery hybrid storage system to the grid-tied photovoltaic system." *IEEE Transactions on Sustainable Energy,* 10.3 (2018): 1370–1379.

[16] Burke, Andrew. "Ultracapacitors: why, how and where is the technology." *Journal of power sources,* 91.1 (2000): 37.

[17] Yang, Yuqing, *et al.* "Battery energy storage system size determination in renewable energy systems: A review." *Renewable and Sustainable Energy Reviews,* 91 (2018): 109–125.

[18] Liu, Xiong, *et al.* "Control of hybrid battery/ultra-capacitor energy storage for stand-alone photovoltaic system." *2010 IEEE Energy Conversion Congress and Exposition, IEEE*, 2010.

[19] Jing, Wenlong, *et al.* "Battery-supercapacitor hybrid energy storage system in standalone DC microgrids: A Review." *IET Renewable Power Generation,* 11.4 (2017): 461–469.

[20] Xu, Qianwen, *et al.* "A decentralized dynamic power sharing strategy for hybrid energy storage system in autonomous DC microgrid." *IEEE Transactions on Industrial Electronics,* 64.7 (2016): 5930–5941.

Power Consumption in Base Station System and Clustered BTS Model

Bablu Kumar Singh[1] and Rajesh Bhadada[2]

Department of Electronics & Communication Engineering, MBM Engineering College, Jai Narain Vyas University, Jodhpur, Rajasthan–342001, India

E-mail: [1]bablukumar.singh@gmail.com; [2]rajesh_bhadada@rediffmail.com

ABSTRACT: *In this paper, the work consists of base stations system (BTS) power consumption with proposed techniques for optimization. This paper also shows the optimization of power at Radio Frequency level using Clustered BTS model for Mobile communication with the help of path loss and compared it with standard model being used in Mobile communication. The energy study showed that mismanagement of lighting systems, air-conditioning systems, and the type of buildings increased the power consumption of the base station. By applying energy savings techniques proposed for the proposed base stations in the upcoming technology for optimized and efficient utilization of power is presented.*

Keywords: BTS, Power Consumption, Mobile Communication, BTS Model, Energy Optimization.

Introduction

Wireless communication has grown immensely throughout the globe, which resulted in growing demands of spectrum and research are being carried out to optimize the existing spectrum to fulfil the growing demands of the spectrum for users. The rapid increase in demand and use of mobile communication resulted scarcity in current spectrum [1], [2]. In order to meet requirements in telecommunication to maximize the range of signals and the extent of the telephone and broadcast coverage, telecommunication and broadcast companies are more focused on energy saving methods to be implemented at BTS which consumes most of the power and due to increasing demands, more and more BTS are being installed which produces additional power consumption and it needs to be addressed because in several rural and urban areas in India (on the mountains and the buildings), these installations require a reliable electric power supply without interruption. According to [3], approximately 3% of the world's electrical energy is consumed by the latest communication technologies specially the hardware equipment's producing approximately 2% of the CO_2 emissions in the world and it is increasing day by day; 9% of this consumption is caused by radio communication networks [4]. Within these radio communication networks, 10% of the energy is consumed by the users of terminals, while 90% is consumed by telecommunication base stations [5]. Thus, the increase in the number of base stations in case of mobile communication increase in the global energy consumption leads to increase in energy costs, which has direct impact on global warming.

The power consumption distributions [6] of the various components of base stations is shown in Figure 1.

It shows that the maximum energy consumer by the radio frequency equipment that includes power amplifier, transceivers and the connecting cables, which consumes approximately 65% of the total energy including air conditioning (17.5%), digital signal processor (10%), and the analog to digital and digital to analog converter (7.5%). In some regions of India temperature are high and to reduce the energy consumption of base stations in these areas air cooling required continuously adding extra load of power consumption. To optimize energy consumption, we need to optimize the power consumption of BTS as well as its components including radio frequency energy. The power consumed at single BTS is listed in Table 1.

Apart from the major equipment various constraints in which Base

Table 1: BTS Equipment Power Consumption [7]

Sl. No.	*Power Consumed at BTS*		
	BTS Equipment	*Characteristics*	*Working hours*
1.	Transmitter/Receiver in GSM 900/ GSM 1800	Max 3600 W × 3 Supply 48 DC	24 h
2.	Two energy bays Ericsson (charger of batteries made of AC/DC rectifier)	Rectifier 220 V–24 V Power 10 KW	24 h
3.	Power supply	Voltage: 220 V/400 V, 10–30 A, $\cos\varphi = 0.8$, Power: 15 kVa	24 h
4.	Power system ericsson	DC output: 24 V or 48 V DC/200 A/6, 4 KW	24 h
5.	Lighting lamps	40 W × 2	12 h
6.	Two equipment radios	Supply 48V DC 2 × 2, 4 W	24 h

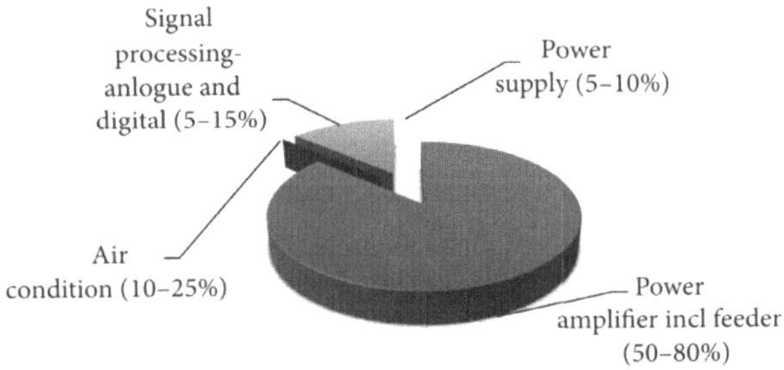

Fig. 1: Energy Consumption by Various BTS Sections [6]

stations equipment are kept consists of cooling system according to environmental conditions i.e., summer and winter seasons has different levels of power consumption. There is also power optimization is possible when different Base Stations work together to form a cluster and can work at Radio Frequency (RF) level to optimize the RF power of the mobile network which is known as spectrum power.

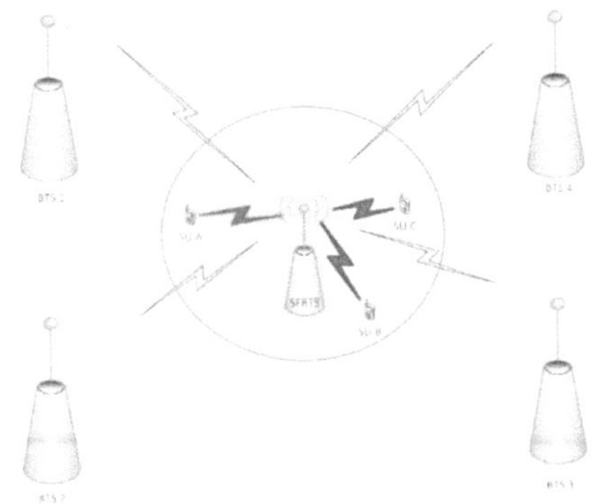

Fig. 2: Clustered - based Station Model

Principle of Cluster based Model

The Clustered based channel model is a modified form of Base Station System (BTS), which keeps track of nearest BTS and make virtual cluster of BTSes. Store Forward plan at Base Transceiver System (SFBTS) within range of few meters to 3 km radius depending upon the type of network shown in Figure 2.

Clustered model of BTS will assumed to be connected to nearby BTS Channel Number (ARFCN) of the mobile service provider. The connection is based on link quality for which path loss needs to be calculated to measure the performance and the quality of the signal when user is connected to nearest base station. Here the spectrum power in terms of frequency channel is being optimized by using cognitive radio technology. The Path loss [1] is different for different technology and needs to be addressed. The designing parameters for Transmitter and receiver for Radio frequency Management we have taken parameters as.

Transmitter Parameter

The design cluster model we have some assumptions that the antenna height taken as 30 m with various parameters shown in Table 2, and by substituting the values Effective Isotropic Radiated Power (EIRP) calculated [8] which gives value 58.5 dBm.

Receiver Parameter

The parameters set for receiver design contains the receiver height at 10 m, taking coverage range of radius 3 KM and standard receiver sensitivity of 114 dBm for general GSM BTS [8]. The maximum path loss to establish link between BTS and SFBTS assuming no polarization loss among transmitter and receiver using Table 2, parameters is 126.54 dBm [8].

Table 2: Transmitter Parameter

Transmitter for SFBTS Model	*Parameter Value*
Power Output	44.5 dBm
Gain of Antenna	18 dB
Losses in Cable	2 dB
Combiner Loss	2 dB
Transmitter antenna height	30 m

According to transmitter and receiver parameter path loss is calculated for clustered model of BTS. The Pathloss calculated has been compared with Okumura model [10] and COST231 model [11]. Using Table 3 parameters, the calculated path loss obtained is 134.48 dBm and 145.1 dBm respectively. The comparison of these two existing channel models with clustered model shows significant improvement in path loss as it is much lower than the existing models for different receiver antenna heights as shown in Figure 3.

Table 3: Receiver Parameter

Parameter	*Specification*
IP Address	192.168.10.3
IQ sampling	200 K
Career Frequency (According to ARFCN)	957.4 MHz
Active Antenna	Rx1
Gain	25dB
Sample per frame	1000

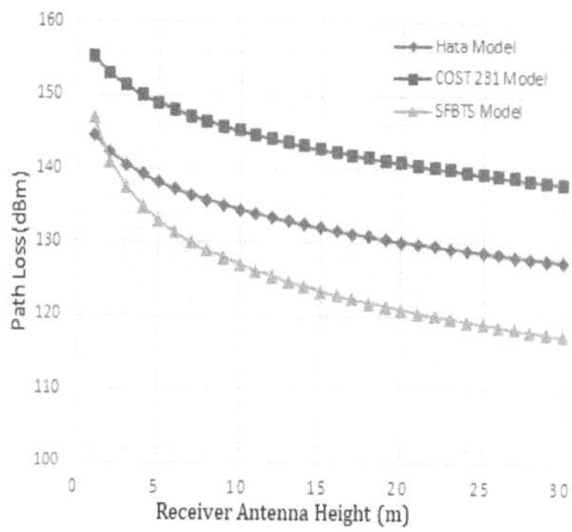

Fig. 3: Path Loss Comparison

BTS Power Optimization Techniques

The technical proposals of energy savings can be applied for BTS power consumption shown in Table 1. For lighting systems: by using more efficient lighting system which will contribute to the reduction of cooling load; For air conditioner system: by maintaining appropriate temperature and humidity inside the base stations cabins and by avoiding air conditioner from direct sunlight or replace it with variable speed fan which may contribute for pawer saving. For base station cabins: it must be seal, windows should be wide enough to utilize advantage of daylight to reduce energy consumption from lighting lamps; Base transmitter station and others equipment: by using energy efficient equipment.

Conclusion

In this paper, we have presented power consumption in BTS and energy saving approaches on the sites of the base stations used for Mobile communication with power optimization approaches to reduce power consumption at BTS. We have also simulated result for Path loss for clustered model of BTS and it has been validated with well-known path loss model used today in mobile communication and the losses occur in our proposed channel model of BTS is 126.54 dBm which is significantly less as compared with existing recommended model shown in Figure 3. The clustered BTS model can be applied as a modified form of BTS for cognitive radio and efficient power utilization at radio frequency level.

Acknowledgment

We express sincere gratitude to the Department of Electronics and Communication Engineering, M.B.M. Engineering College, Jai Narain Vyas University, Jodhpur, Rajasthan, India for providing access to research lab which enabled to complete this work.

References

[1] Rashid, Rozeha A.; Sarijari, Mohd Adib; Fisal, Norsheila; Lo, Anthony; Yusof, Kamilah S. and Mahalin, Nurhija, "Spectrum Sensing Measurement using GNU Radio and USRP Software Radio Platform," in *Seventh International Conference on Wireless and Mobile Communications (ICWMC),* 2011.

[2] Dutta, U.K., Razzaque, M.A., Al-Wadud, M. Abdullah; Islam, M.S., Hossain, M.S. and Gupta, B.B., "Self-Adaptive Scheduling of Base Transceiver Stations in Green 5G Networks," *IEEE Access*, Vol. 6, 2018, pp. 7958–7969.

[3] Humar, I., Ge, X., Xiang, L., Jo, M., Chen, M. and Zhang, J., "Rethinking energy efficiency models of cellular networks with embodied energy," *IEEE Network*, Vol. 25, No. 2, pp. 40–49, 2011.

[4] Ericsson, "Sustainable energy use in mobile communications," *Ericsson White Paper,* 2007.

[5] Vignolles, Dennis, Report TIC Development Durable, 2008. http://www.ladocumentationfrancaise.fr/var/storage/rapports-publics/094000118.pdf.

[6] Study Energy Efficient Radio Access Network Technologies, "Alcatel-lucent/TU Dresden Vodafone chair mobile communications systems," *ATIS Report on Wireless Network Energy Efficiency,* 2009.

[7] Ayang, Albert; Ekam, Paul; Videme, Bossou and Temga, Jean (2016). Power Consumption: Base Stations of Telecommunication in Sahel Zone of Cameroon: Typology Based on the Power Consumption-Model and Energy Savings. *Journal of Energy,* 2016, pp. 1–15.

[8] Singh, B.K. and Bhadada, R., "Enhancing Spectral Efficiency of Mobile Communication by Store Forward Base Transceiver System,"

in 7th *International Conference on Signal Processing and Integrated Networks (SPIN),* Noida, India, 2020, pp. 781–784.

[9] Intuitive guide to principles of communications. Available at: http://complextoreal.com/wp-content/uploads/2013/01/linkbud.pdf.

[10] Farhoud, M., El-Keyi, A. and Sultan, A., "Empirical correction of the Okumura-Hata model for the 900 MHz band in Egypt," *Third International Conference on Communications and Information Technology (ICCIT),* Beirut, 2013, pp. 386–390.

[11] Yang, Yun; Zhang, Jie and Wang, X., "GSM network optimization: Design and implementation of an algorithm dealing with path propagation loss," *4th International Conference on Computer Science & Education*, 2009, pp. 307–311.

Effect of Temperature and Irradiance on Grid Connected Solar PV System

Danish Khan[1] and Dhanesh Kumar Sambariya[2]

Department of Electrical Engineering, University College of Engineering, Kota, Rajasthan Technical University, Akelgarh, Kota, Rajasthan–324010, India

E-mail: [1]danishpiet506@poornima.org; [2]dksambariya@rtu.ac.in

ABSTRACT: *Non-conventional energy resources such as solar, wind, tidal provide clean energy. As the power demand increases rapidly, power failures occurs in the system. Renewable (nonconventional) energy sources are used to provide the power for constant load and fulfil the demand continuously. PV is the most-easiest and efficient source to generate power but due to variations in the environmental condition, the energy conversion is low. To increase the energy conversion, MPPT technique has been used, which detects the peak power to produce the maximum energy. In this paper, study done on the working of grid-tied solar PV array with the effect of irradiance and temperature change. The components used in the system and analysis of the result of the grid station or load is performed. This paper studies a 120 KW grid-tied solar PV array, MATLAB simulated model and the effects the power generation by solar PV with variation in irradiance and temperature.*

Keywords: PV Array, MPPT Technique, Boost Converter, IGBT Inverter, Transformer, Grid.

Introduction

Energy is the most beautiful gift to human beings. The consumption of energy shows the growth of any society or country. Worldwide energy consumption is increasing day by day and the energy demand has been largely met by fossil fuel and owing to their exhausting nature the fossil fuel are reducing at a very fast rate. It is estimated that 80% of fossil fuels will be over by 2050 [1]. Due to the heavy consumption of fossil fuels, we are facing environmental problems like pollution and global warming. Nuclear energy is the partial solution to this problem but by the using of nuclear energy is limited because of safety concerns and growing opposition by the local population. Due to the problems of the undesirable release of contaminants and the exhaustible nature of fossil fuel, it is necessary to find alternatives that are inexhaustible and pollution-free. These never-lasting sources also commonly known as renewable energy sources or non-conventional energy sources are more effective and offer many advantages over conventional energy sources. Increasing negative effects on the environment and the ever rising energy need of people lead to the popularization of these renewable resources [2].

Indian government always try to encourage peoples for move to RE (Renewable energy) and plant Non-conventional energy projects in various states of India. The solar plant is the most famous generation resource to generate energy by RES. India Meteorological Department (IMD) maintaining a network of radiation stations and they measure the sunshine duration of the day. Rajasthan has tons of land which is barren and very little population so as these areas are suitable for large central power stations based on solar energy. In the Rajasthan amount of solar radiation it received throughout of the day on clear sunny days is 2543°C temperature and 230–800 W/m^2 radiation [3–6]. In the introduced model help to users to know to generate electricity through solar panel built upon the buildings' roof & parking, and tied with the grid. It will help to reduce the user's bills and the earning if power is supplied to the grid also.

The PV has used to convert sunlight directly in the DC power. The conversion of the energy is called the photoelectric effect. There are generally 2 types of the PV Arrays, these arrays are two types polycrystalline or monocrystalline and change to solar energy into electrical energy and it changes of output power depends on the temperature and solar irradiation. So to obtain maximum power at the result end side boosting the voltage by control the current of array by VSC controller. By depending on the boost converter the output AC voltage may be change and connected to utility Grid or to the load of various applications. The main objective of this paper to study a detailed model of 120 KW generated by PV system and connected to the grid or utility loads then observe the performance evaluation of this system VSC controller.

Model Formation and Structure

Grid-connected PV systems may be ranges from relatively small units, from installation on the rooftop, for distributed power generation; to big centralized power generation plants. This system takes in Solar panels, power electronic converters which include the dc-ac inverter, boost converter, maximum power point tracking (MPPT) controller, loads at Point of common coupling (PCC), and sometimes filter components.

The schematic diagram for a grid integrated system feeding on a solar photovoltaic has been shown in Figure 1. By using a boost converter, improve solar panel output voltage so it has to match with DC-link voltage, which has to be controlled appropriately to produce a smooth voltage waveform at grid level by the inverter. The performance of this system will be regulated by the way in which the inverter is get connected. There is a lot of inverter configuration available which has to be connected between solar PV and Grid, and it is strongly recommended to select a suitable configuration for getting optimum results from the system.

Multilevel Inverter

A multilevel inverter is a power electronic scheme that extracts a required output voltage from a number of levels of dc voltages used as inputs. It consists of an array of power semiconductor devices, clamping diodes and capacitor voltage sources. As the number of dc sources is increased, the voltage waveform gets more steps & hence the output voltage coming from the inverter, approaches towards nearly sine waveform while using a fundamental frequency switching scheme. The most advantage of this type of converter is that it produces a stepped voltage waveform with small steps, resulting in better output quality, lower harmonics, lower switching losses, higher efficiency & better electromagnetic computability. A multilevel inverter has one more advantage in that it not only achieves high power ratings, but also enables the integration of renewable energy sources, such as photovoltaic, wind, and fuel cells can also be interfaced in a very easy manner for a high power application.

As in the block diagram it shows how energy is transfer and convert in different forms. We generate the DC power by using PV system then convert the DC power into AC by using Inverter and transferred to the transformer for increase voltage and match to the grid voltage. The PV system has no energy storage losses since there are no battery used as it's not a standalone system.

Photovoltaic Array

A solar photovoltaic (SPV) cell generates electric energy when a light beam/sunlight falls on it. And this effect is known as the photovoltaic effect, observed initially in 1877 in a solid material. These cells are made up of semiconductor materials, two types are available: n-type and p-type. The n-type has electrons majority in the conduction band, while the other one yields a whole majority in the valance band. In this way, PV cell has a p-n junction, if incoming photons contain more energy than the bandgap, electrons pass from valance band to conduction band through this junction and thus the electron-hole pair has formed [7–8].

Solar PV cell shows a non-linear current and voltage characteristics and it's totally depends on solar radiation and temperature.

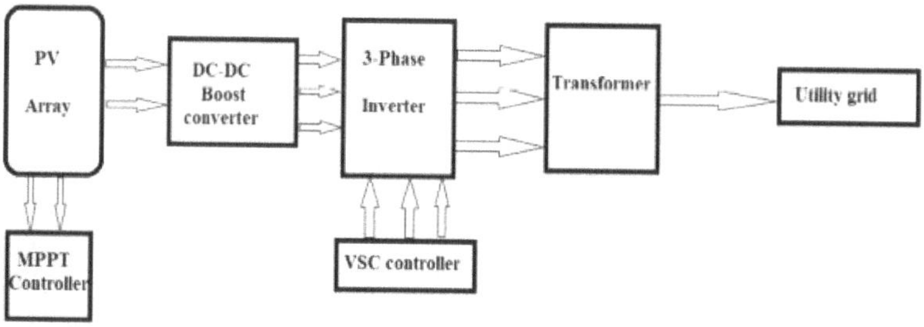

Fig. 1: Block Diagram

$$I = I_L - I_s\left(\frac{V}{e^{n.V_T}} - 1\right)$$

Here I_L is photo current, $I_s = I_d$ is saturation current of diode, n is diode ideality factor,

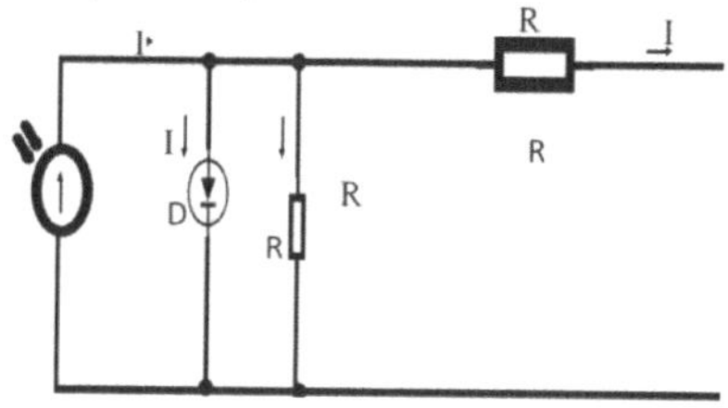

Fig. 2: Solar Cell Equivalent Circuit

MPPT Controller

With Sun location and direction of the rays are changes throughout the day and the power output also changes through the PV module. So to ensure the effectual operation of the solar array the maximum power point of the array has to be tracked. MPPT, which is an addition to rising power given from the PV module to the load, is considered as a PV system lifetime booster [5]. Maximum power point (MPP) does not lie at a point but it moves around the PV curve depending on the radiation and temperature [11–12].

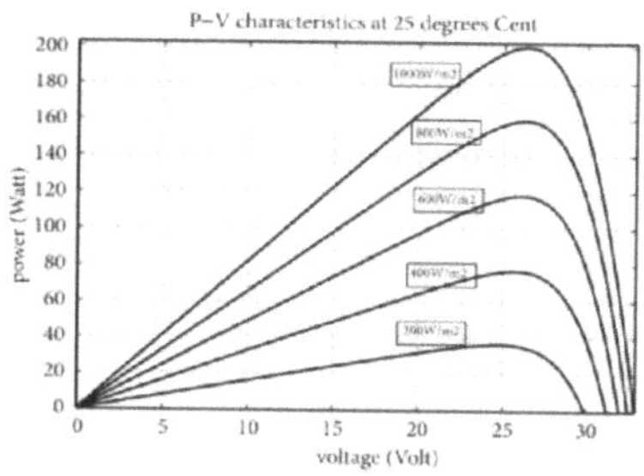

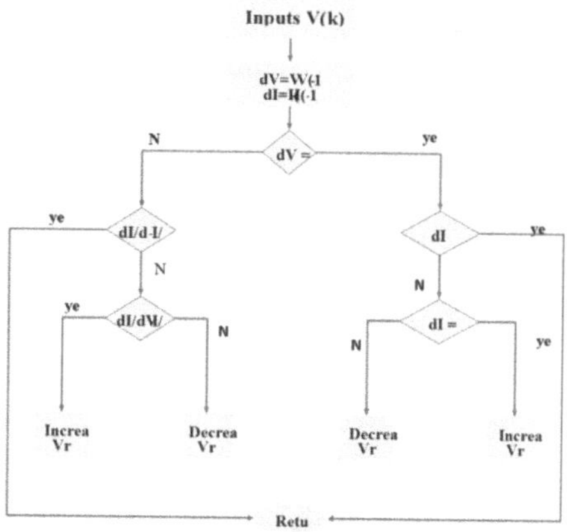

Fig. 3: State-flow Chart of IC MPPT Technique

DC-DC Boost Converter

A DC-DC boost converter is used to enhance the level of output voltage. Here a DC-DC converter using the IGBT (insulated gate bipolar transistor)

technique the high frequency square waveform is used for switching operation of IGBT. And due to the negative connection to the diode, it acts as a high impedance path. Here no output is appearing across the capacitor after switching so voltage has to add $V_{dc} + V_1$. Where V_1 is induced voltage from the inductor since load get the voltage, as well as a capacitor, starts charging when IGBT get enabled then capacitor and give power to the inverter. An inverter is responsible to convert DC power to AC power.

Transformer

A transformer has to be used to gain sustained voltage which is required for supply. In the PV system transformers are added or used to change the ratio of voltage to current delivered from the voltage source to meet the characteristics of the load to keep the power constants. The primary coil of the transformer receives energy from inverter, a secondary coil that received energy from the primary winding and delivered it to the load or the grid and to the core which makes a path for the magnetic flux. The input voltage at the primary side is 260 V and the transformer steps up the voltage up to 25 KV at the secondary side and after it, transmit the voltage to the utility grid where after passing from the feeder the voltage will again step up and input voltage at the primary side is 25 KV and step up output voltage is 120 kV.

Model

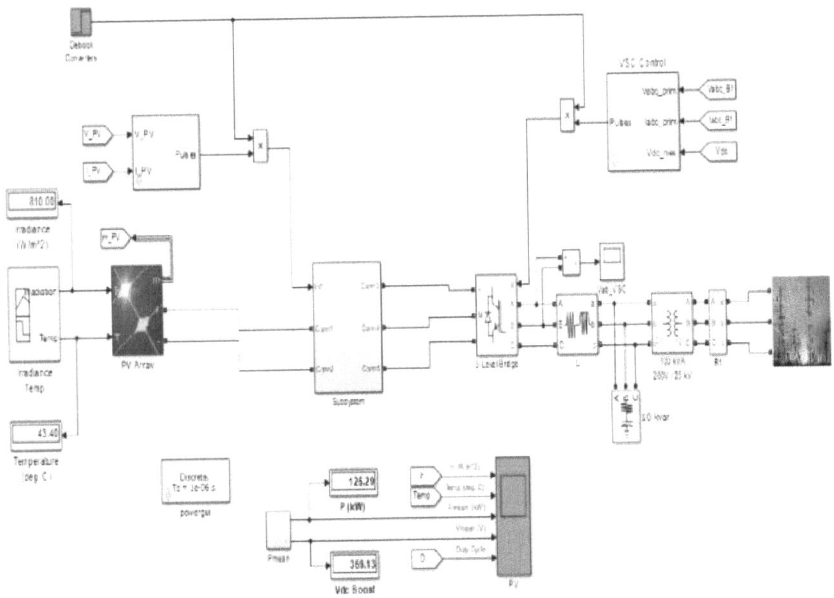

Fig. 4: Simulation Model of 120 KW Grid-tied PV System

Results

This model show the performance of the grid tied solar PV system with the change in irradiance and the temperature. Figure 5 shows the output of the DC-DC converter, where peak voltage is 400 V. The change in the irradiance from 800 to 250 W/m^2 at the time of 0.6 to 1.1 sec. is considered and observed the changes in the output power from 120 KW to 50 KW for the same time. After that change in irradiance from 250 to 800 W/m^2 at the time 1.2 sec. to 1.7 sec and observed that power increases from 50 KW to 120 KW according to irradiancc.

In waveform 5(b) change in temperature from 25°C to 43°C at the time from 2 to 3 sec. and observed there are slightly change in the output power the power is decreased. In Figure 6 show the output of transformer, the output voltage is 20 KV and the current decrease as the energy transmit from primary to secondary coil remain constant.

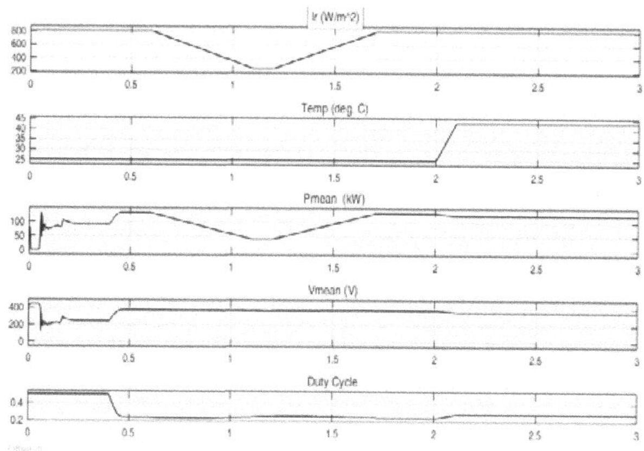

Fig. 5: Output of DC-DC Converter

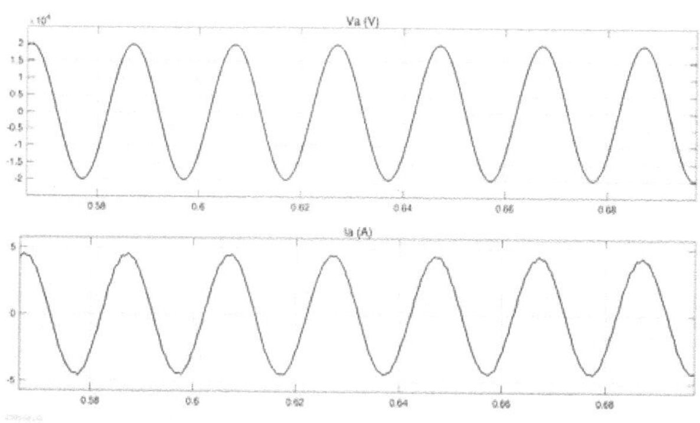

Fig. 6: Output from the Transformer

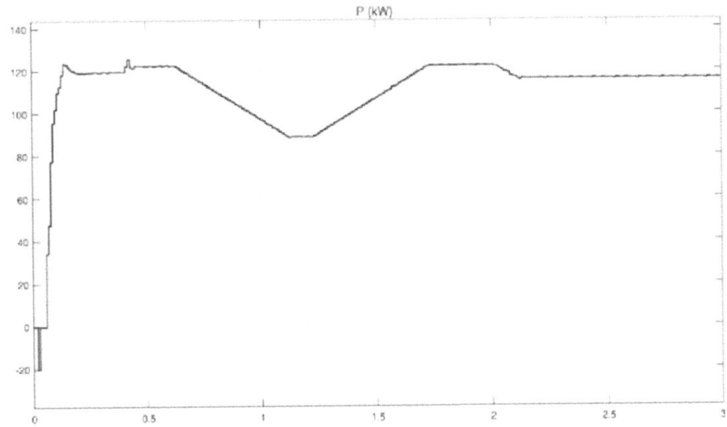

Fig. 7: Power at Grid

Conclusion

In the last few years, a huge transformation has been observed concerning the electrical sources used in the power system. The distribution generation is attending more concentration as compared to centralized power generation because several non-conventional energy sources can be employed at DGs resulting in more and more benefits especially for the concerns related to the environment. Nowadays, the recent advancements and continuous research in PV technology have led to a large number of grid integrated solar PV systems along with their use in an offgrid system where the power transmission from utility systems is a difficult task. In this research, a thorough study and analysis of a grid-tied PV system have been carried out with giving major concentration on the improvement of power quality and the variation of output power with the change of irradiation and temperature.

In this paper a MATLAB Simulink model is constructed a detailed representation of PV system. MATLAB simulation of 120 KW solar PV grid connected system and observe the performance evaluation of the system with variation of the temperature and the irradiation. And study about the changes in the generating power with the changes of the temperature and the radiation. Solar PV system is take as a primary source three phase inverter has to use to convert DC source to AC source in this type of voltage source controller the main advantage is output waveform not depend on the type of load.

References

[1] Xiao, W., Edwin, F.F., Spagnuolo, G. and Jatskevich, J., "Efficient Approaches for Modeling and Simulating Photovoltaic Power Systems", *IEEE Journal of*

Photovoltaics, Vol. 3, No. 1, pp. 500–508, Jan. 2013.

[2] Yazdani, A., Fazio, A.R.D., Strunz, K. *et al.*, "Modelling Guidelines and a Benchmark for Power System Simulation Studies of Three-Phase Single-Stage PV systems," *IEEE Transaction on Power Delivery,* Vol. 26, No. 2, pp. 1247–1264, Apr. 2011.

[3] Muirhead, Ian and Hawkins, Barry, "Research into new technology photovoltaic modules at Telstra Research Laboratories——What we have learnt", 1996.

[4] Ankala, Burri, R. Sharadha jalakanuru nageswararao implementation of mppt algorithm for solar photovoltaic cell bycomparing shortcircuit and incremental conductance method.

[5] Jimenez, J.F., Ortiz, D. Biel Solé. Study and simulation of photovoltaic systems-electrical conversion using Matlab/Simulink, 2009.

[6] ECR Solar. Photovoltaic Energy Book. Chapter 4: The Photovoltaic Panel, pp. 27–35.

[7] Reisi, A.R., Moradi, M.H., Jamasb, Sh., "Classification and comparison of maximum power point tracking techniques for photovoltaic system: a review", 2013. *Energy Rev.,* 19(C), 433–443.

[8] Mohanta, Prasanta Kumar. Modeling and Control of a Grid Connected Photovoltaic System. Diss. 2016.

[9] Kumar, B. Shiva and Sudhakar, K., "Performance evaluation of 10 MW grid connected solar photovoltaic power plant in India." *Energy Reports,* 1, 184–192, 2015.

[10] Bharathkumar, M., and Byregowda, H.V., "Performance Evaluation of 5 MW Grid Connected Solar Photovoltaic Power Plant Established in Karnataka." *International Journal of Innovative Research in Science, Engineering and Technology,* 3.6, 2014

[11] Ahmad, Zameer, and Singh, S.N.. "Modeling and Control of Grid Connected Photovoltaic System—A Review." *International Journal of Emerging Technology and Advanced Engineering,* 3.3, 2250–2459, 2013.

[12] Faranda, Roberto and Leva, Sonia. "A Comparative Study of MPPT techniques for PV Systems." *7th WSEAS International Conference on Application of Electrical Engineering (AEE'08),* Trondheim, Norway. 2008.

[13] Lamnadi, M., *et al.* "comparative study of ic, p&o and flc method of mppt algorithm for grid connected pv module." *Journal of Theoretical & Applied Information Technology,* 89.1, 2016.

[14] Atiq, Jawairia and Soori, Prashant Kumar. "Modelling of a Grid Connected Solar PV System Using MATLAB/Simulink." *International Journal of Simulation—Systems, Science & Technology*, 17.41, 2016.

[15] AbdelHady, Rameen. "Modeling and simulation of a micro grid-connected solar PV system." *Water Science,* 31.1 1–10, 2017.

[16] Dhaked, Dheeraj Kumar and Birla, Dinesh, "Microgrid Designing for Electrical Two-Wheeler Charging Station Supported by Solar PV and Fuel Cell", *Indian Journal of Science and Technology*, 14(30):2517–2525, 2021.

[17] Dhaked, Dheeraj Kumar and Birl, Dinesh, "Battery Charging Optimization of Solar Energy based Telecom Sites in India", *Engineering, Technology & Applied Science Research*, Vol. 9, No. 6, 5041–5046, 2019.

Electric Process Heater and Heater Control Methodology

Divya Vigneshraja

Electrical & Instrumentation Engineer (Cameron Manufacturing India Pvt. Ltd.), 3rd Floor, Tidel Park, Vilankurichi Road, Coimbatore–641014, India

E-mail: divyananthakumar.n@gmail.com

ABSTRACT: *Electric process heaters are used in all the major industries ranging from Onshore to Offshore, Power Generation, Metal, Polymers etc., Control of these Process Heaters, to achieve required heating and switching shall be achieved through Electronic devices like thyristors and contactors or combination of both. The appropriate switching of the Heater, control over the required range of temperature and cut-off/tripping of the heater on fault shall be achieved through the proper scheming and wiring of the electronic switching devices and the protection devices. This Paper shall demonstrate the details of Process Heater and its control philosophy using Thyristor Bank.*

Keywords: Electric Heater; Thyristor; Oil & Gas; Control Panel.

Introduction

In Oil & Gas industry, treating & conditioning the extracted Oil, Gas or water is a major process. Electrical Process heaters are used to heat process gases or fluids to maintain precise process temperature control and to attain various process conditions. Electrical heating is based on the principle, that when electric current passes through a medium, heat is produced. Let us take the case of solid material which as resistance 'R' ohms and current flowing through it is I amps for 't' seconds than heat produced in the material will be $H = I^2Rt$. There are various methods for controlling the temperature supplied to the Heater elements and this shall be varied by varying the supply current.

Electric Heating

There are many advantages of using Electricity for Heating over the conventional methods and a few important advantages are listed below:

1. High Factor of Safety
2. No Pollution
3. Smaller Footprint
4. Faster response
5. Low Cost

6. Precise Control
7. Instant Cold Start

There shall be immersion type (Tank/Vessel) heating requirement or inline type (Pipe) heating requirement based on the Process conditions.

Heater Construction

- ***Heating Elements:*** Heating Coil shall be made of Nickel Chromium NiCr (80–20) resistive wire in a tube compacted with Magnesium oxide powder which provides Electrical Insulation & Thermal conduction. Design of the heater elements shall ensure minimum peak inrush current.
- ***Tube Plate:*** The Heating elements shall be connected to the tube sheath either by bite coupling or welding. Selection of materials shall be suitable in preventing cracking caused by stress corrosion. Water ingress into the heater element shall be prevented by means of suitable provision material and method.
- ***Enclosure/Vessel:*** The whole arrangement of heating elements and tube sheath will be placed inside a vessel.
- ***Baffles & Tie Rods:*** To hold the heating elements and for appropriate distribution of heat, baffles & tie rods are used.
- ***Terminal Box:*** The Heating elements are connected to the terminal box and cables will be connected to the terminal box through the cable glands. The maximum allowable temperature in the terminal box (up to 65°C maximum) and the internal space of the heater with restricted flow shall be the base factor in determining the required unheated element length.

Instruments: For over temperature protection of the Heater, 2 Nos. of heat sensing devices (Thermocouple, RTD, Temperature transmitters etc.) shall be connected to the heating elements and 1 No. of the sensing device to the Tube sheath. On over temperature, the signals from these instruments, trips the heaters and cuts the power supply to the Heater. Figure 1

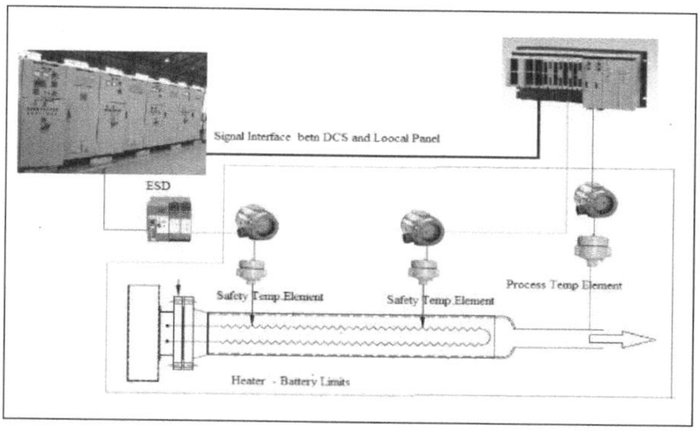

Fig. 1

shows the protection scheme of Heater using instrument and control system.

Heater Control Panels (HCP)

The major element of HCP is the Medium Voltage 2 Leg Thyristor controller with built in semiconductor fuse for the protection of the controller. The HCP shall be protected by a 6.6 kV Vacuum Circuit breaker (VCB) and a 6.6 kV Vacuum Contactor (VC). Due to the frequent switching nature of the Heater, Vacuum Contactor is used and due to the high current application a VCB is also considered. This system is a combination of VCB + VC protected circuit. A Lighting arrestor shall be provided for surge protection in the incoming side of the HCP and a combination of Core Balance current transformer with Earth Leakage relay is considered on the outgoing side of the HCP. Current Transformers (CT) shall be provided for Protection relays & Metering systems, Potential Transformer for Voltmeter shall be provided. The control power supply required for various application inside the HCP is derived internally by means of a control transformer. Human Machine interface shall be achieved by means of Emergency Push Buttons, selector switches & Annunciator panel. A few indication lamps for indicating the HCP and Breaker conditions, power supply indications & Heater conditions shall be provided. Figure 2 shows the sample Single Line diagram of MV Heater control panel.

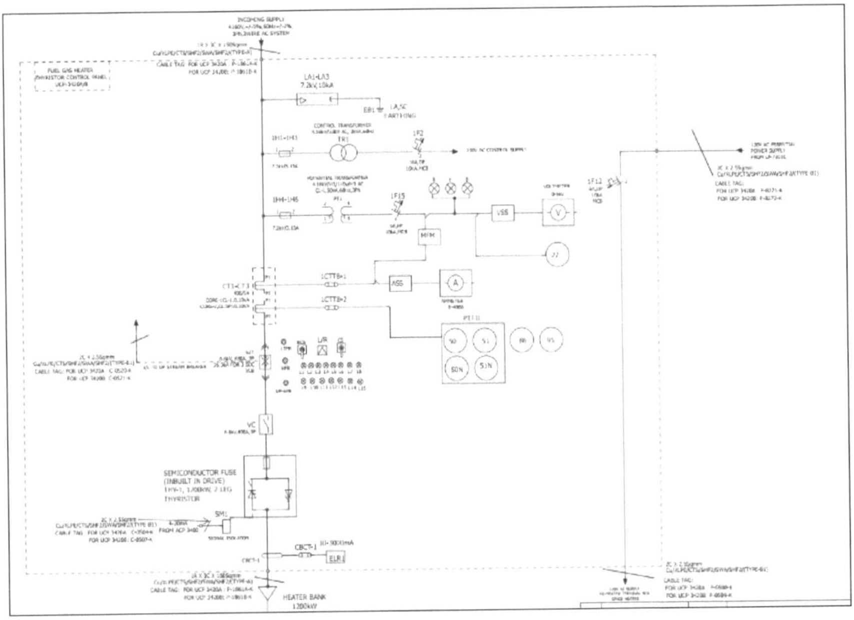

Fig. 2

Design Requirements

Heater and Control assembly shall be designed to ensure safety and reliability. The heat flux requirement depends on the heated medium and it is measured in Watt/cm^2.

Heater shall be capable of operating with allowable tolerance limits on supply voltage and frequency. The control panel for heaters shall be appropriately screened to optimize electromagnetic compatibility (EMC). The heater control panel shall have an adequate level of intrinsic immunity to external electromagnetic disturbance to enable it to operate as intended. Electromagnetic disturbance generated by the heater control panel and its individual components shall not exceed a level which would affect the correct operation of either radio or telecommunications equipment. Maximum temperature of the heating element shall not cause degradation of process fluid/Gas.

Control and Protection Methodology

Heater's power shall be controlled by thyristors fired according to the zero-crossing mode, with a controlled output power range. By varying the firing angle to the thyristor bank, the power input to the Heater is controlled. Based on the temperature measured on the heated output of the heater, the firing is so adjusted to reach the required temperature.

The temperature from the MV heater is sensed using temperature sensors fixed to the Heater Elements (2 Nos.) and to the Heater Flange (1 No.). Once the sensed temperature crosses the setpoint, an over-temperature trip signal is triggered, and it trips the Breaker of the HCP. Heater shall be switched from either in local mode or through the soft signals in Remote mode. Upon fault in the Heater/HCP, a trip signal will be sent to the associated MV Switchgear breaker, to isolate the faulty feeder from the system. Equipment protection shall be as follows:

- Overcurrent protection for thyristors e.g. ultra rapid fuses, voltage transient suppressers, di/dt limiting reactors, dv/dt protection networks etc.
- Over-temperature protection with pre-alarm and trip signal. The connection can be either connected directly to the power and control assembly's main contactor or via an instrument interface.
- Over-voltage protection to avoid damage to any analogue instruments from faults in the thyristor trigger circuit.

Protective relaying shall include over current, under voltage, Earth leakage & Master trip. Indications shall include Breaker Open, Close & Trip, Heater Bank On & Off, Local/Remote mode, Supply voltage On (RYB).

Feedbacks shall include Heater shutdown, Breaker position & Emergency stop.

Space heater shall be provided for Heater terminal box and Heater control panel to avoid moisture accumulation/ condensation. These anti-condensation heaters shall be automated with the pre-set Thermostats.

Inspection and Testing

Testing of the Heater and the control assembly shall be complete with Visual, Routine & Functional testing.

Routine tests include High Voltage (10 kV for 1 min.) test, Insulation resistance test before and after the HV test.

Functional tests include operational testing of VCB, VC, Thyristor bank, relays, Meters, selector switches & Annunciator Panel. Checking of the auxiliary power supplies to the panel and interlock & trip simulation shall be done.

References

[1] IEC 61892 – Mobile and fixed off-shore units – Electrical installations.

[2] IEC 61000 – Electromagnetic compatibility (EMC) IEC 60947 – Low-voltage switchgear and control gear.

Synchronized Detection and Discrimination of External and Internal Shunt Faults in FSCTL Using Two Zones Current Information

Shoyab Ali[1] and Gaurav Kapoor[2]

[1]Department of Electrical Engineering,
Vedant College of Engineering and Technology,
Tulsi, Bundi–323021, Rajasthan, India

[2]Department of Electrical Engineering, Modi Institute of Technology,
Nayagaon, Rawatbhata Road, Kota–324010, Rajasthan, India

E-mail: [2]gaurav.kapoor019@gmail.com

ABSTRACT: *This paper presents a protection system based on wavelet analysis (WA) for the synchronized detection and discrimination of external and internal faults in FSCTL (fixed series compensated transmission line) using two zones current information. The system detects the faults occurring inside the protected area and outside the protected area synchronously. A two-area power system network is considered and it is simulated using MATLAB software. The simulation results validate that the protection system is correctly detecting and discriminating the external and internal faults below a semi-cycle instant.*

Keywords: Fault Detection; Fixed Series Compensated Transmission Line; Fourier Transform; Shunt Faults; Wavelet Transform; Zones Protection.

Introduction

The most researched areas of the power system protection are the fault identification and its localization. On the other hand, the analysis of power system and research has inflicted enormous significance on its relaying schemes. Researchers have been executing on the various schemes of fault analysis for TLs since a little decades before. The authors in [1] have addressed the classification of faults in TLs by using principal component analysis (PCA). In [2], the authors present experimental results and discussions which verify that their introduced methodology based on optimized ensemble of regression tree locates the evolving faults in TLs correctly. Further, a comparative study of the presented fault identification and localization method using local current has shown that the former one is more accurate than the latter [3]. The authors of [4] have introduced an application of signal entropy in fault analysis, whereas a robust protection scheme is proposed in [5] using convolutional neural network. Supervised PCA is used in [6] to develop effective fault diagnosis scheme. Probabilistic generative model is also used in this field as a separate method of fault analysis [7]. Empirical mode decomposition is used as a hybrid method with random forest classifier and modified impedance technique [8] in this field of research with outstanding results. Some of the researches have also executed a bunch of techniques, like PNN (probabilistic neural network) [9] and ANN (artificial neural network) with wide area management systems [10] for fault recognition in TLs.

This paper introduces synchronized detection and discrimination of external and internal shunt faults in FSCTL utilizing two zones current information. These currents are of fundamental frequency attained by applying Fourier transform. Further the current samples are decomposed using wavelet analysis to get the approximate coefficients. Finally, the difference of approximate coefficients (DOAC) is calculated for the synchronized detection of external and internal faults.

Explanation of Two Zone System

A two-zone system is used in this work for all simulation studies. MATLAB software is used to model the system. The schematic diagram of the modeled system is illustrated in Figure 1. The two power generation plants are connected through a 300 km FSCTL of 400 kV. The synchronous measurement of 3-Φ currents is done at MB-1 (measurement bus-1) and MB-2 (measurement bus-2). The area between MB-1 and MB-2 is selected for simulating internal 1-Φ-G, Φ-Φ, and Φ-Φ-G faults. The area beyond plant-2 is used for simulating external 1-Φ-G, Φ-Φ, and Φ-Φ-G faults.

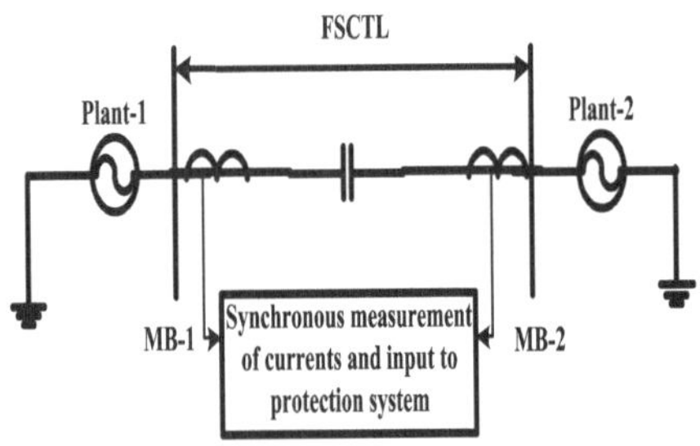

Fig. 1: Schematic of the Test System

The specified proposed protection system is presented by flowchart as shown in Figure 2. The current samples are obtained from MB-1 and MB-2 synchronously and are given as input data to the protection system. The selection of threshold value (THV) was made considering simulations for different 1-Φ-G, Φ-Φ, and Φ-Φ-G type external and internal faults for different fault inception time.

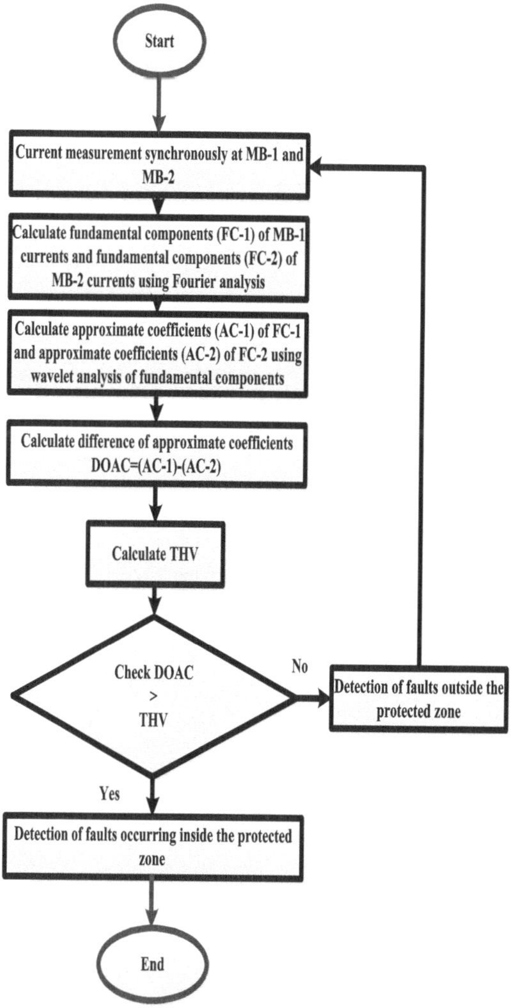

Fig. 2: Flowchart of the Protection System

Results and Discussion

A systematic simulation was performed on MATLAB for presenting the efficacy of the protection system. For all the cases, the sampling frequency was sustained at 1 kHz. The results are given below.

Line to Line and Line to Line to Ground Internal Faults and Line to Ground External Faults

The reliability of the protection system has been analysed for LLG fault in ΦA-ΦB-G at 0.1 s inside the protected

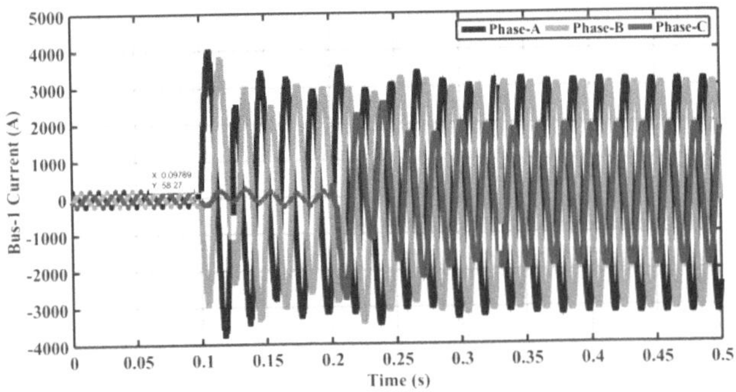

Fig. 3: Measurement Bus-1 Currents during Internal Fault ABG at 0.1 s and External Fault CG at 0.2 s

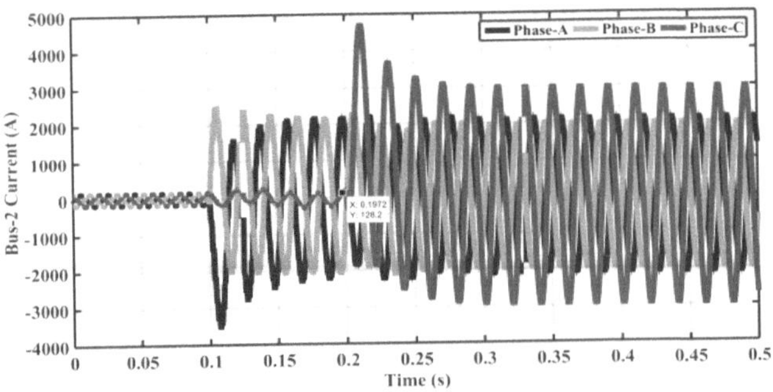

Fig. 4: Measurement Bus-2 Currents during Internal Fault ABG at 0.1 s and External Fault CG at 0.2 s

area and LG fault in ΦC at 0.2 s outside the protected area. Figure 3 presents the 3-Φ current waveform measured at the measurement Bus-1 with internal fault in ΦA and ΦB and external fault in ΦC. Figure 4 shows the 3-Φ current waveform measured at the measurement Bus-2 with internal fault in ΦA and ΦB and external fault in ΦC. The discrimination of internal and external faults has been carried out on the basis of two THVs i.e. +THV and –THV respectively. Figure 5 demonstrates the plot of synchronized detection and discrimination of internal fault ABG when detected at 5 ms and external fault CG when detected at 5 ms. The difference of approximate coefficients during various LLG faults inside the protected zone and LG faults outside the protected zone is presented in Table 1. The detection time of fault during LLG faults inside the protected zone and LG faults outside the protected zone is presented in Table 2. It is evident from Table 1 and Table 2 that synchronized detection and discrimination of LLG type internal faults and LG type external faults has been successfully conducted.

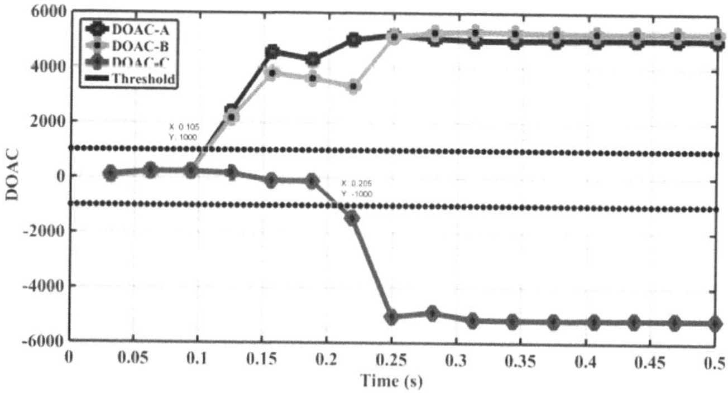

Fig. 5: Synchronized Detection and Discrimination of Internal Fault ABG at 5 ms and External Fault CG at 5 ms

Table 1: Difference of Approximate Coefficients during LLG Faults Inside Protected Zone and LG Faults Outside Protected Zone

LLG Fault Inside Protected Zone, FIT = 0.1 s	*LG Fault Outside Protected Zone, FIT = 0.2 s*	*Difference of Approximate Coefficients*		
		DOAC of ΦA	*DOAC of ΦB*	*DOAC of ΦC*
ABG	CG	5.1977×10^3	5.3318×10^3	223.9377
AC	BG	3.6663×10^3	224.9213	4.9038×10^3
BCG	AG	222.8236	5.3473×10^3	5.3510×10^3
AB	CG	4.9789×10^3	3.6348×10^3	223.7249
ACG	BG	5.2916×10^3	222.7879	5.0768×10^3

Table 2: Detection Time of Fault during LLG Faults Inside Protected Zone and LG Faults Outside Protected Zone

LLG Fault Inside Protected Zone, FIT = 0.1 s	*LG Fault Outside Protected Zone, FIT = 0.2 s*	*Detection Time of Fault (ms)*		
		DT of ΦA	*DT of ΦB*	*DT of ΦC*
ABG	CG	5	5	5
AC	BG	5	5	10
BCG	AG	10	10	5
AB	CG	5	5	10
ACG	BG	5	5	10

Line to Ground Internal Faults and Line to Line and Line to Line to Ground External Faults

The reliability has been analysed for LG fault in ΦA at 0.2 s inside the protected area and LLG fault in ΦB-ΦC-G at 0.3 s outside the protected area. Figure 6 presents the 3-Φ current waveform measured at the measurement Bus-1 with internal fault in ΦA and external fault in ΦB and ΦC. Figure 7 shows the 3-Φ current waveform measured at the measurement Bus-2 with internal fault in ΦA and external fault in ΦB and ΦC. The discrimination of internal and external faults has been carried out on the basis of two THVs i.e. +THV and –THV respectively. Figure 8 demonstrates the plot of synchronized detection and discrimination of internal fault AG when detected at 10 ms and external fault BCG when detected at 15 ms. The difference of approximate coefficients during various LG faults inside the protected zone and LLG faults outside the protected zone is presented in Table 3. The DOTF during LG faults inside the protected zone and LLG faults outside the protected zone is presented in Table 4. It is evident from Table 3 and Table 4 that synchronized detection and discrimination of LG type internal faults and LLG type external faults has been effectively conducted.

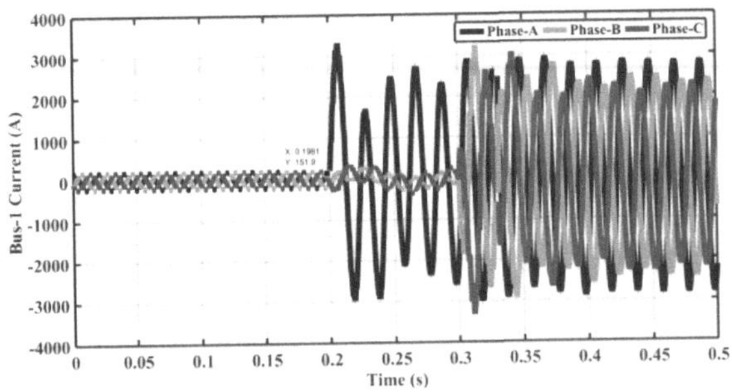

Fig. 6: Measurement Bus-1 Currents during Internal Fault AG at 0.2 s and External Fault BCG at 0.3 s

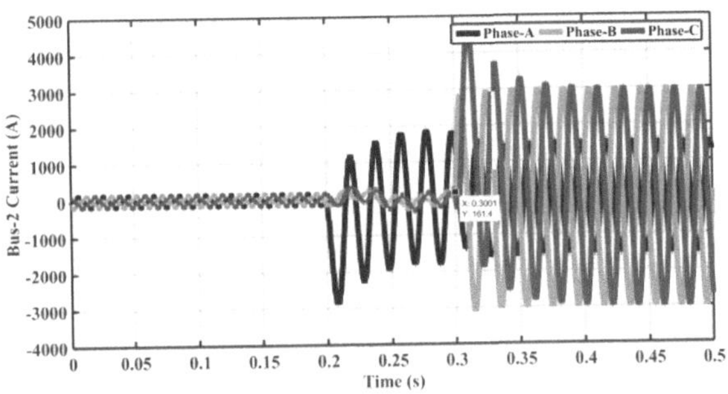

Fig. 7: Measurement Bus-2 Currents during Internal Fault AG at 0.2 s and External Fault BCG at 0.3 s

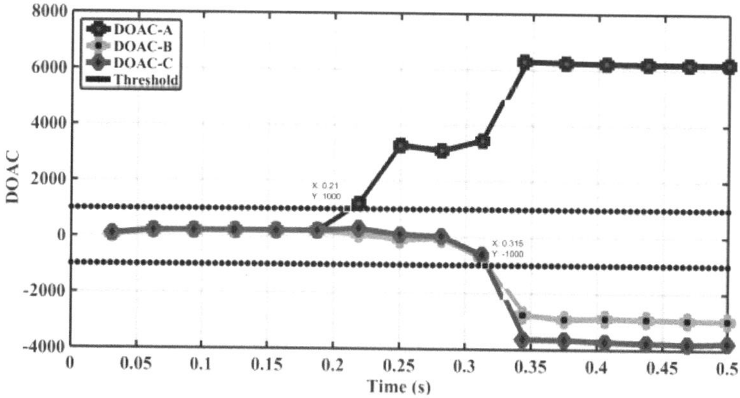

Fig. 8: Synchronized Detection and Discrimination of Internal Fault AG at 10 ms and External Fault BCG at 15 ms

Table 3: Difference of Approximate Coefficients during LG Faults Inside Protected Zone and LLG Faults Outside Protected Zone

LG Fault Inside Protected Zone, FIT = 0.2 s	*LLG Fault Outside Protected Zone, FIT = 0.3 s*	*Difference of Approximate Coefficients*		
		DOAC of ΦA	*DOAC of ΦB*	*DOAC of ΦC*
AG	BCG	$6.3018*10^3$	222.7931	303.1376
BG	AC	250.7824	$3.0386*10^3$	223.9687
CG	ABG	222.9241	272.4162	$6.2339*10^3$
BG	ACG	251.1611	$6.2167*10^3$	223.9377
AG	BC	$3.2795*10^3$	222.8836	303.0968

Table 4: Detection Time of Fault during LG Faults Inside Protected Zone and LLG Faults Outside Protected Zone

LG Fault Inside Protected Zone, FIT = 0.2 s	*LLG Fault Outside Protected Zone, FIT = 0.3 s*	*Detection Time of Fault (ms)*		
		DT of ΦA	*DT of ΦB*	*DT of ΦC*
AG	BCG	10	15	15
BG	AC	15	20	20
CG	ABG	15	10	10
BG	ACG	10	15	15
AG	BC	10	25	25

Conclusion

In this paper, an efficient protection system has been introduced for FSCTL. The synchronized detection and discrimination of external and internal shunt faults in FSCTL using two zones current information has been performed accurately using WA. The simulation results prove that the protection system is able to detect and discriminate the external and internal faults synchronously in below a semi cycle instant.

References

[1] Mukherjee, A., Kundu, P.K. and Das, A., "A Differential Signal-Based Fault Classification Scheme Using PCA for Long Transmission Lines", *Journal of The Institution of Engineers India Series B*, Vol. 102, pp. 403–414, 2021.

[2] Valabhoju, A., Yadav, A., Pazoki, M. and Sehiemy, R.A., "Optimized ensemble of regression tree-based location of evolving faults in dual-circuit line", *Neural Computing and Applications*, Vol. 33, pp. 8795–8820, 2021.

[3] Elmitwally, A. and Ghanem, A., "Local current-based method for fault identification and location on series capacitor-compensated transmission line with different configurations", *International Journal of Electrical Power and Energy Systems*, Vol. 133, 107283, ISSN 0142-0615, 2021.

[4] Mukherjee, A., Kundu, P.K. and Das, A., "Classification and Fast Detection of Transmission Line Faults Using Signal Entropy", *Journal of The Institution of Engineers India Series B,* Vol. 102, pp. 655–670, 2021.

[5] Tong, H., Qiu, R.C., Zhang, D., Yang, H., Ding, Q. and Shi, X., "Detection and classification of transmission line transient faults based on graph convolutional neural network," in *CSEE Journal of Power and Energy Systems*, Vol. 7, No. 3, pp. 456–471, 2021.

[6] Mukherjee, A., Kundu, P.K. and Das, A., "A supervised principal component analysis-based approach of fault localization in transmission lines for single line to ground faults", *Electrical Engineering*, Vol. 103, pp. 2113–2126, 2021.

[7] Fahim, S.R., Sarker, S.K., Das, S.K., Islam, M.R., Kouzani, A.Z. and Mahmud, M.A.P., "A Probabilistic Generative Model for Fault Analysis of a Transmission Line With SFCL," in *IEEE Transactions on Applied Superconductivity*, Vol. 31, No. 8, pp. 1–5, 2021.

[8] Biswas, S. and Nayak, P.K., "A New Approach for Protecting TCSC Compensated Transmission Lines Connected to DFIG-Based Wind Farm," in *IEEE Transactions on Industrial Informatics*, Vol. 17, No. 8, pp. 5282–5291, 2021.

[9] Mukherjee, A., Chatterjee, K., Kundu, P.K. and Das, A., "Probabilistic Neural Network-Aided Fast Classification of Transmission Line Faults Using Differencing of Current Signal", *Journal of The Institution of Engineers India Series B*, Vol. 102, pp. 1019–1032, 2021.

[10] Barreto, N.E.M., Rodrigues, R., Schumacher, R., *et al.*, "Artificial Neural Network Approach for Fault Detection and Identification in Power Systems with Wide Area Measurement Systems", *Journal of Control, Automation and Electrical Systems*, 2021.

Application Based Rapid Prototype Development using TMS320F28335 DSP and MATLAB/Simulink

Gopal Lal Jat*, Vijay Sirohi, Jagdish Kumar and Shiv Narayan
Department of Electrical Engineering, Punjab Engineering College,
Chandigarh, Sector 12, UP–160012, India
E-mail: *gopallaljat.phdele@pec.edu.in

ABSTRACT: *Now-a-days, engineers are constantly facing challenges to do programing in microcontroller based software-hardware to design a real time control system. The programing process into any microcontroller takes a lot of time and causes delay in the prototyping. In the modern era of rapidly growing technology, developers want to make brisk prototyping process. This article aims at developing a hardware for rapid real time electrical applications. The MATLAB/Simulink environment provides a swift and fast process for programming and rapid prototyping of hardware based applications. There are various methods available to make hardware and software interfacing with graphical or model based programing. Texas Instruments™ (TI) and MATLAB™ have developed such environment where real-time Micro Controller Unit (MCU) can be efficiently programmed using Simulink blocks. The MATLAB embedded coder and Code Composer Studio (CCS) configure and build the code. This code flashes to Digital Signal Processor (DSP) and its variants. In this paper a 3-phase Induction Motor is interfaced with the DSP using power converter in the MATLAB environment.*

Keywords: CCS; DSP; MATLAB/Simulink; Rapid Prototyping; Motor Control.

Introduction

The abrupt evolution of power electronics over the past decades is mainly an outcome of two factors. The prime factor being development of high switching frequency semiconductor switches while second one is an advancement and integration of real time complex controllers. These controllers bridge the gap between hardware and software computations. The complex algorithms find their way in hardware through the use of soft computing. Thus it offers cost effectiveness, user friendly nature and flexibility in controlled converters. Also, these converters find their application in UPFCs, STATCOMs and various distribution generation schemes.

The input line current consists of dominant harmonics which are not desired at all. These harmonics are removed using PWM converters which also uses a small passive filter with an aim of removing the harmonics of higher order [1–2]. The size, weight and response of PWM rectifier is more impressive than conventional rectifiers [3]. The precision and performance are the concerning parameters that evolved with digital control using DSP. The DSP controller offers the advantages of reliability and programmability. Moreover, DSP is flexible to adapt to various control strategies depending upon the application of interest while keeping the same hardware intact to it. DSP can process high speed control algorithm in real time. All of the above mentioned characteristics of DSP make it advantageous to have digital control of 3 phase and 2 level voltage source converter. In this article, interfacing of DSP TMS320F28335 has been implemented using MATLAB, thus bringing down the effort and equipment required in pulse generation. These pulses can be programmed in MATLAB.

For DSP TMS320F28335, design of closed loop feedback with the PWM techniques is very complex and takes more time in CCS. Therefore, MATLAB embedded coder generates the code for real time implementation. MATLAB is very impressive tool for DSP controller as it has inherent advantages like generation of code and simple design. Also, debugging is easy and can be accomplished for controlling steps [4]. The extensive objective of this article is to give simple and clear steps to program the TMSS320F28335. Firstly, embedded coder, Target Support Package (TSP) and Simulink coder are configured with CCS.

Automated code generation tool employed to initialize, configure and generate C code from Simulink environment. This code is flashed to DSP for the following tasks in sequence: real time simulation, embedded system deployment and rapid prototyping.

The generation and execution of stand-alone C codes for testing and developing algorithms replicated in Simulink is done by real time workshop. The resulting code obtained from hardware and Simulink model is applicable to real time as well as non-real time applications. The applications include acceleration of simulation, hardware-in-the-loop testing and rapid prototyping. Target template files are queued by real time workshop to transfer Simulink models into C code [5] [7]. The environment for running the code is specified in the target templates. User specified targets and third-party targets, ready-to-run configurations can be developed keeping the constraint that the target should be supported by Real-Time Workshop (RTW). To favour this process, Embedded Target for TI TMS320C2000 DSP [6] dwells TI C2000 target (C2000lib block set) that boosts the automatic rapid prototyping on a C2000 hardware.

Embedded Target for TI TMS320 C2000 DSP is used for the integration of Simulink and MATLAB with Texas Instruments Embedded Coder Support Package tools. The development tools of TI pick the C code. This C code is produced with real-time workshop [8]. The code is customized according to hardware target. The hardware is further supported for Embedded Target for TI TMS320C2000 DSP. The development tool also builds an executable file for the processor. Moreover, the build options in Real-Time Workshop constructs a studio project for code composer. Thus every attribute featured by CCS assists in developing the desired algorithm or application. After downloading the executable files, they are run on the hardware. The compilation process of these files can be accessed from two sources only i.e. CCS or MATLAB. Further, MATLAB requires two powerful tools to access the running code i.e. Link for code composer studio [9] and Real-Time Data Exchange.

Simulation and System Design

The simulation is in accordance to control algorithm and specifications of the hardware involved. The induction motor control setup is first modelled into Simulink and then implemented into real time hardware setup. The instrument tool of MATLAB i.e. TSP consists of various default modules for user to understand and program [10]. Some of the modules are described below in Table 1.

There are six PWM (ePWM) modules which are independent of each other in TMS320F28335 and have the capability to generate complex waveforms. The ePWM has two output channels namely ePWMA and ePWMB. Both of these channels belong to the same ePWMx module. Further every channel and its module comprises of seven different sub modules to realize various functions in the generation of PWM waveforms. In practice, the waveform is generated by comparing two waveforms i.e. carrier and modulated waveform. Carrier wave

Table 1: TI C28335 DSP Target Supported Library Module in Simulink

Module	*Purpose*
C28x-ePWM	PWM generation module to generate high frequency PWM signal for convertor
C28x-eCAP	Sense rising and falling edge of pin connected signal. It gives the speed information, position and direction of motor
C28x-GPIO DI	Read digital input (0 to 5 V) signal like limit switch to user interface
C28x-GPIO DO	Generate digital output (0 to 5 V) signal to operate relays or other LED
C28x-SCI	Data exchange from DSP to PC via serial communications interface (RS-232)

is generated by DSP timer initialization and the modulated wave is a control signal generated from controller command. Timer value is obtained from a Timer Base Counter Register (TBCR) and reference value is obtained from a set point. The set point is stored as Compare Register-x (CMPx) value in DSP memory. The PWM signal is generated based on option selected into ePWMx module.

The PWM frequency is changed by selection of Time Based Period Register (TBPRD) in module. Equation (1) is gives desired PWM frequency of gate signal.

$$f_{PWM} = \frac{1}{2 * TBPRD * T_{TBCLK}} \quad \dots (1)$$

Where, T_{TBCLK} is time-base clock period that can be equal to system clock period if clock division is selected as 1.

$$TBPRD = \frac{1}{2} * \frac{f_{SYSCLKOUT}}{f_{PWM}} \quad \dots (2)$$

The system clock frequency ($f_{SYSCLKOUT}$) of the DSP F28335 is 150 MHz and as the design considerations of this project, gate pulse frequency is selected as 10 KHz. So, from the above relation between PWM frequency and time base clock period, the value of TBPRD can be calculated as below.

$$TBPRD = \frac{1}{2} * \frac{150*10^6}{10*10^3} = 7500 \quad \dots (3)$$

The CMPx register is related to PWM duty cycle (D) by the formula:

$$CMPx = (1 - D) \times TBPRD \quad \dots (4)$$

For example, consider half duty cycle (D = 0.5) then CMPA value is calculate by equation 7.

$$CMPx = \frac{TBPRD}{2} = 3750 \quad \dots (5)$$

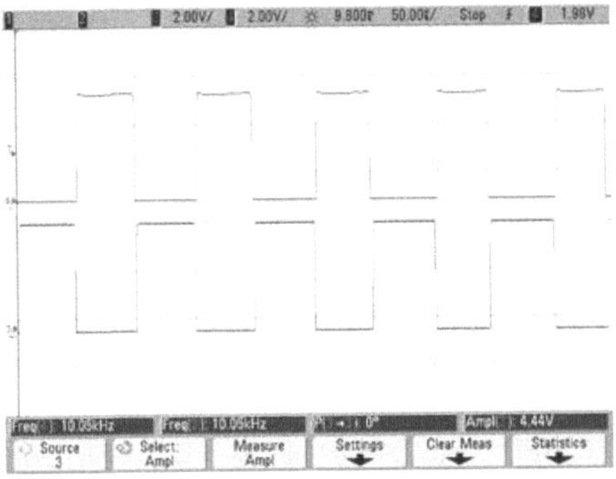

Fig. 1: Complementary PWM Pulses Generation by ePWM Module

The dead band between two complimentary pulses of same leg IGBTs is introduced from ePWM module. The sine PWM can be generated using DSP by selecting a sinusoidal signal as a reference and triangular wave as a carrier signal. The basic parameter about PWM frequency and modulation can be adjusted in ePWM module.

The capture (eCAP) module is able to count rising or falling edge of input signal in predefine time. So if any pulse signal having variable frequency is applied to its pin, it can measure frequency of signal. In practical application the encoder having 500 pulses per revolution is connected to motor shaft such that it gives 12.8 KHz frequency signal when revolving at 1500 rpm. QEPA and QEPB are the two quadrature signals from encoder having 90° phase shifted. Now the DSP need to measure the frequency of input signal and convert it into rotations per minute. The QEPA signal has a frequency which is captured by eCAP module and stored as capture count. This capture count is number of rising edge in a particular time. Thus, speed can be accurately measured by given formula

$$\text{Motor speed (RPM)} = \frac{17578125}{\text{Capture count}} \quad \dots (6)$$

The user interface is developed by General Purpose I/O (GPIO) pins on DSP. The reference speed can be adjusted by GPIO push button for increment and decrement set speed. The user is also able to regulate speed from host personal computer.

The MATLAB model is networked with hardware to do the real-time interfacing. The interfacing and modelling is done according to control structure and machinery is arranged in such a manner that it controls the speed of three phase induction motor.

Hardware Implementation

The hardware setup is designed for closed loop control of three phase motor control using variable frequency method. The hardware implementation requires basic power convertor like ac to dc convertor for making a constant dc link voltage across the inverter input terminal and again this dc link voltage is converted into variable ac using dc to ac convertor. The controlled signal for dc to ac convertor is gate pulses that may be generated by various PWM method.

In this paper sine PWM method is used to reduced harmonics distortions on output side of convertor. The Eddy current loader is coupled with ac motor to loading motor up to a rated level. The DSP now play controller role, as it stores the *C* code which burns from MATLAB embedded coder and control the whole closed loop performance of induction motor. The proportional integral and derivative (PID) tune the closed loop and generate appropriate modulation index value for sine reference signal to maintain certain output voltage/frequency ratio in various conditions. The block diagram for induction motor closed loop speed control using DSP controller is shown in Figure 2.

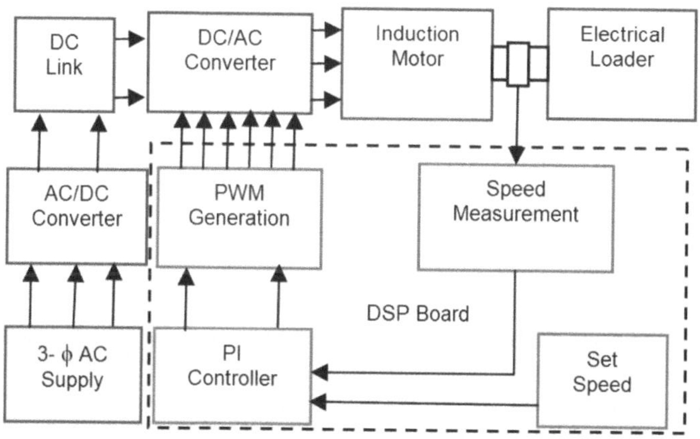

Fig. 2: Block Diagram for Speed Control of Motor using DSP Setup

Table 2: Hardware Specification for Motor Control

S. No.	*Hardware*	*Specification*
1.	Induction Motor	3-ϕ, 0.75 kW, 415 V, 1410 rpm, 50 Hz ac input
2.	DSP	TMS320F28335
3.	Invertor	18 kW, 3-ϕ 4 leg Semikron kit
4.	Loader	dc motor with resistive load

The C code is flashed on DSP board using JTAG emulator. The level shifter is amplifying gate pulses to high voltage level so as to trigger power IGBTs. Gate signals are also optically isolated using isolator board to interface between high voltage and low voltage.

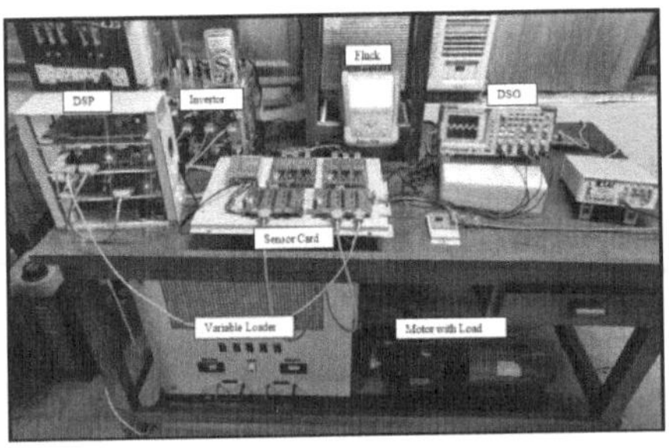

Fig. 3: Hardware Implementation of DSP based Motor Drive System

The eddy current loader is an electric loader that is coupled with induction motor. The dc injector current controls the loading. The current is driven by a variable dc regulator system. The torque sensor based on strain gauge senses the torque on shaft. This torque based on strain force is displayed in seven segment. The host receives the real time speed, torque and current values of motor through RS 232 cable. The same real time data transmitted from DSP is plotted (stored to workspace) into MATLAB/SIMULINK using SCI receive module.

Experimental Results

The V/F ratio is presented in Table 3 and it is nearly constant for a large range over which motor is operated. The GPIO switch on DSP board gives the 'set' command to step and ramp input. The hall-effect sensor is used to sense the current and the value is measured by DSP (using ADC module).

Table 3: Evaluation of V/F Ratio at Various Speeds

Speed (RPM)	*Voltage (Volt)*	*Frequency (Hz)*	*V/F Ratio*
398	64.9	12.83	4.83
602	94.8	19.8	4.72
803	125	26.9	4.59
997	150.9	33.5	4.49
1204	176.1	40.2	4.33
1401	201.0	47.4	4.22

All the data accumulated is transmitted to PC (MATLAB) using SCI module. The step and ramp response of motor speed in real time at no load and full load are shown in given Figure.

Table 4 summarizes the time response analysis of step input. The results conclude that the speed response of motor at no load and full load has dynamic response.

Similarly, the ramp response on no load and full load are shown in Figure

5. The slope of ramp input requires pre-adjusted and is done using Simulink. The acceleration of ramp signal is set to 100 rpm per sec.

The motor speed does not trace the set speed until 300 rpm due to flux weakening in the low frequency region. After the low frequency region, the speed traces the set speed with a small drift value. The full load and no load steady state error are 5.0% and 4% respectively.

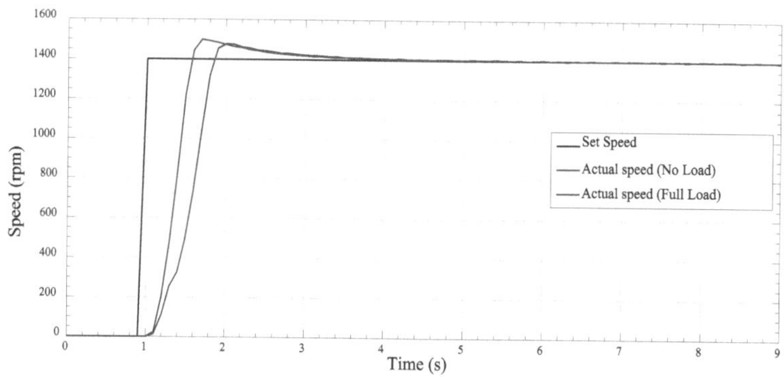

Fig. 4: Real Time Speed Step Response of Induction Motor

Table 4: Time Response Analysis of Step Speed Response

	Delay Time (s)	*Rise Time (s)*	*Peak Time (s)*	*Settling Time (s)*	*Over Shoot (%)*
No Load	00.42	00.31	00.71	2.40	7.07
Full Load	0.62	0.46	1.08	2.53	5.56

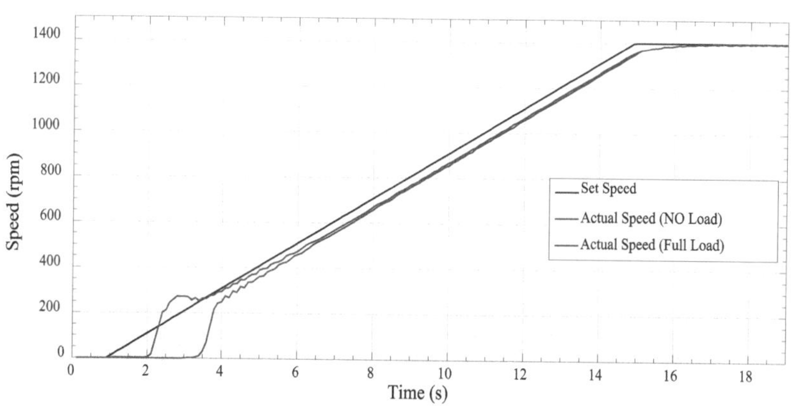

Fig. 5: Real Time Speed Ramp Response of Induction Motor

The limitation of this *V/F* control technique that is it is unable to trace the input below 280–300 rpm. Thus, the control technique used in this paper is favourable for all the applications where desired speed of motor is above 25–30% of rated speed.

Conclusion

This paper presents the rapid prototyping process with the help of MATLAB/Simulink. The main idea is to overcome the false fear of researchers regarding programming into C and product prototype. MATLAB have various option to build program without knowledge of C or C++. The MATLAB/Simulink provides an efficient way of using embedded coder tools. The user does not write codes for these tools which directly impacts the time saving. The process is highly efficient and is faster in execution. Thus, it becomes attractive for industrial control, design and research. The real time approach is expressed into steps with fully real time standalone hardware development. The model for speed control of induction motor is designed. This model is implemented using DSP. The DSP processor is used to produce Sinusoidal PWM (SPWM) signals. The programming of a dead band is done for a three phase operation. The output from dead band is flashed to inverter module. The DSP based system is conclusively fast, user friendly and cheap. It also reduces the complexity by a large margin. The results from the experiment reveal that control schemes based on DSP have higher computing power and better performance than the conventional control techniques. Thus speed control is most effective with DSP. The beauty of Simulink is that it has inbuilt blocks and tools which are often required for any power electronic, motor drive and smart grid application.

References

[1] Bocker, Joachim and Mathapati, Shashidhar, "State of the Art of Induction Motor-Control", *IEEE Conference on Electric Machines & Drives, Antalya*, Vol. 2, pp. 1459–1464, May 2007.

[2] Ling, L. and Mekhilef, S. "DSP Based Implementation of Adaptive Speed Controller for Three-phase Induction Motor", *IEEE international conference,* Vol. 2, 2009.

[3] Hur, B., "Learning Embedded Systems with MSP432 Microcontrollers: MSP432P401R with Code Composer Studio", 2nd ed.; Independently Published: College Station, TX, USA, 2020.

[4] Prabaharan. V. and Umamaheswaran, R, "Microcontroller based Speed Control of Sinusoidal PWM Inverter fed Three Phase Induction Motor," *International Journal of Innovative Technology* and Research, Vol. 1, Issue 1, pp. 061–065, December–January 2013.

[5] Chaudhari, P.S., *et al.*, "Comparison of Performance Characteristic of Squirrel Cage Induction Motor by Three Phase Sinusoidal and PWM Inverter Supply using MATLAB Digital Simulation," *IEEE Third International Conference on Emerging Trends in Engineering and Technology*, pp. 362–367, 2010.

[6] Sathishkumar, *et al.*, "Microcontroller Based BLDC Motor Drive for Commercial Applications in Power Electronics and Renewable Energy Systems", *Lecture Notes in Electrical Engineering;* Springer: New Delhi, Vol. 326, pp. 829–841, India, 2015.

[7] Gopal Lal Jat, *et al.*, "Real time speed control of induction motor using new generation DSP controller", In *Proceedings of the 2016 Second International Innovative Applications of Computational Intelligence on Power, Energy and Controls with their Impact on Humanity (CIPECH),* pp. 90–95, Ghaziabad, India, 18–19 November 2016.

[8] French, C.D., *et al.*, "Rapid prototyping of a real time DSP based motor drive controller using Simulink", *International Conference on Simulation,* pp. 284–291, York, UK, 1998.

[9] Jamadar, B., *et al.*, "Design and Development of Control System for Three Phase Induction Motor using PIC Microcontroller", In *Proceedings of the 3rd International Conference on Advances in Control and Optimization of Dynamical Systems*, pp. 807–811, Kanpur, India, 13–15 March 2014.

[10] Mikhael, H., *et al.*, "Speed Control of Induction Motor using PI and V/F Scalar Vector Controllers", *International Journals of Computer Application,* Vol. 151, pp. 36–43, 2016.

[11] Narayanan, G. and Ranganathan, V.T., "Triangle-comparison approach and space vector approach to pulsewidth modulation in inverter fed drives", *Journal of the Indian Institute of Science*, Vol. 80, pp. 409–427, Sep./Oct. 2000.

[12] Hong-Ming Chen, *et al.*, "Electro-hydraulic position servo system Sliding mode control Varying boundary layer", *International Journal of Advanced Manufacturing Technology*, Vol. 26, pp. 117–123.

[13] Texas Instrument TMS320x28xx, 28xxx DSP Peripheral Reference Guide (SPRU566E), October 2007.

Effective Cooling Strategy for Solar PV Panels for the Climatic Conditions of Jodhpur

Harish Kumar Khyani[1*], Jayashri Vajpai[1] and Rajendra Karwa[2]

[1]Department of Electrical Engineering, M.B.M. Engineering College, J.N.V. University, Jodhpur–342011, India

[2]Department of Mechanical Engineering, M.B.M. Engineering College, J.N.V. University, Jodhpur–342011, India

E-mail: *khyani.harish@gmail.com

ABSTRACT: *Jodhpur is a promising location for generation of electricity by solar photovoltaic (PV) cells because of the abundance of sunny days with high solar insolation. However, high summer temperature raises the maximum operating temperature of PV panels to as high as 60°–70°C. This rise in temperature severely affects its efficiency and hence its power output, also causing the development of thermal stresses that may reduce their life span. Hence, the cooling of the PV panels to decrease their operating temperature is essential. This paper presents a brief study of cooling methods of PV panels. The use of phase change material (PCM) has been reported to be the most effective among the various cooling techniques employed, because the phase-change process (melting) imparts a high heat storage capacity to PCM. A detailed discussion on the selection of PCM has been presented by considering the ambient temperature conditions of Jodhpur in the summers.*

Keywords: Solar Energy; Photovoltaic (PV) Cells; Active and Passive Cooling; Phase Change Materials (PCM).

Introduction

The conventional electric power generation is associated with high carbon dioxide emissions with high global warming threat and this emphasises the need for greater use of solar PV systems. Jodhpur, located in the thar desert of Rajasthan, is popularly known as the Suncity of India because it is blessed with abundance of sunny days with high solar insolation. Hence, it is making rapid progress in the use of solar PV generation systems. The installed capacity of solar PV plants in Jodhpur has increased significantly in last decade, from 71 MW in 2011–12 to 3831.55 MW in 2021–22 (till 31.8.21) [1].

The data of mean daily maximum and minimum temperatures along with the peak maximum and minimum temperatures over the century for Jodhpur is given in Table 1 [2]. The temperature in the summer months can be seen to be quite high. Since, the solar plants are directly exposed to the sun radiation, their temperatures are prone to be much larger than the ambient temperatures.

The output power and efficiency of PV panels are significantly governed by the operating temperature of the cell. The increase in cell temperature leads to thermal stresses in the panels and reduces their life span. Hence, the cooling of panels to check the impact of the rise of cell temperature on the performance of solar PV panels is essential for their performance enhancement. In the last decade, many investigations have been carried out to control the PV cell temperature using passive and active techniques of cooling.

The objective of this paper is to present a brief note on the effect of cell

Table 1: Mean Daily Maximum and Minimum Temperatures along with the Peak Maximum and Minimum Recorded Temperatures of Jodhpur for the Period of 1901 to 2010 [2]

Month	*Mean Daily Min. Temp. (°C)*	*Mean Daily Max. Temp. (°C)*	*Highest Max. Ever Recorded*		*Lowest Min. Ever Recorded*	
			Temp. (°C)	*Date*	*Temp. (°C)*	*Date*
January	8.9	25.4	35.0	28, 91	–2.2	31, 05
February	11.4	28.1	38.3	28, 53	–0.7	07, 74
March	17.1	33.8	42.5	30, 84	4.7	09, 79
April	22.7	38.9	48.0	25, 58	9.4	07, 18
May	26.6	41.5	48.9	25, 32	17.0	07, 82
June	28.0	40.2	48.0	04, 91	18.8	03, 97
July	26.7	36.1	45.6	04, 01	16.2	28, 2k3
August	25.4	34.2	42.9	05, 57	19.4	15, 37
September	24.2	35.5	42.8	11, 15	15.4	11, 15
October	19.8	36.4	42.3	02, 68	10.0	31, 49
November	14.5	31.8	38.4	01, 86	4.8	25, 93
December	10.3	26.9	34.8	04, k8	0.6	23, 45

temperature on the performance of PV panels, followed by the study of cooling methods to identify the most effective among them. Finally, a detailed discussion on the selection of PCM has been presented by considering the ambient temperature conditions of Jodhpur in summers.

Effect of Rise of Cell Temperature on Performance

Among the numerous influencing factors, temperature has a major impact on the output power and efficiency of a PV cell. Photovoltaic panels convert a part of the incident solar energy into electricity and dissipate the rest as heat. The absorbed radiation leads to increase of operating temperature of the cells. Hence, power output and efficiency reduce due to increased internal carrier concentration and recombination rates.

In comparison with the values at standard test conditions (STC), the open circuit voltage decreases by about 0.1–0.3 V/°K (approximately 0.37%) while, the short circuit current increases by only about 2.3 to 4 mA/°K (approximately 0.05%) for crystalline silicon cells, consequently decreasing the power output substantially at a rate of about 0.4 to 0.5%/°K [3, 4]. The effect of temperature on the output power is depicted in Figure 1. The effect of high temperature of a location leads to high operating temperature of the cell.

The quantitative analysis of the rise of cell temperature by considering climatic conditions at a location is important. The measurement of solar panel temperature was experimentally carried out on the solar panel installed in Electrical Engineering Department, J.N.V. University, Jodhpur using Infrared Thermometer (AR320). It has been observed that, in the month of April, May and June 2021, the temperature of the solar panel reaches up to 58.6°C, 65.7°C and 63.9°C respectively, on the front side.

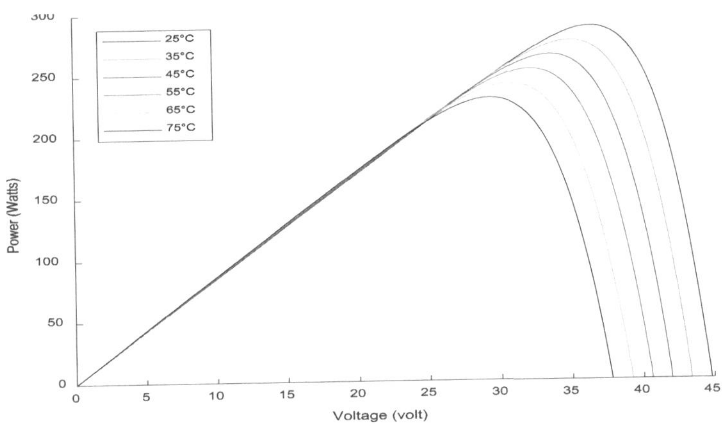

Fig. 1: Effect of Temperature on the P-V Characteristics [5]

As already discussed, the highest ambient temperature in Jodhpur during May is 41.5°C and the cell temperature reaches up to 65.7°C as recorded. As per the STC, the cell temperature should be 25°C for maximum power output. So, ideally the solar PV panel must be cooled by about 40°C in the peak summer conditions of Jodhpur.

Since the temperature coefficient for maximum power is 0.44%/°C [6], the estimated losses will be 17.91%/°C leading to estimated total power losses in Jodhpur, with the current installed capacity of 3831.55 MW, to be 686.23 MW. If an appropriate cooling method is used to reduce the temperature of the cells, then this massive loss can be arrested. Even by reducing the cell temperature from 65.7°C to 40°C, this loss will be reduced by 63%, which is equivalent to a plant of approximately 430 MW capacity.

Cooling Techniques

A variety of passive and active cooling techniques have been suggested in the literature [7]. These are:

1. ***Sensible Cooling:*** By using air draft or a low density, high thermal conductivity and thermal capacity material on the back of the panel.
2. ***Use of Fins:*** Use of fins on the back to enhance heat rejection rate from the back of the panel by increasing the heat transfer area.
3. ***Evaporative Cooling:*** It is an effective way of cooling accomplished by either of the following two methods or by adopting both of these methods.
   (a) Controlled spray of water on the front with or without wipers (to avoid contamination of the glass surface).
   (b) Controlled spray of water on the back of the PV panel.

   The drawback of this method is the requirement of salt free water (especially when water is sprayed on the front of the panel), a reservoir to store water, a small pump to pressurize the water and nozzles to spray water. This scheme draws a part of the developed power of the cell hence it is a costly system and regular maintenance will be required.
4. ***Use of Phase Change Material (PCM):*** Unlike evaporative cooling, it is passive method, i.e., requiring no power. A small quantity of the PCM is capable of removing large quantity of heat because the melting PCM draws latent heat, which is quite significant as compared to the sensible heat. Further, the process of melting is an isothermal (constant temperature) process and the heat transfer coefficient of the melting substance is also quite significant. In general, passive cooling using PCMs have been found to be very effective and provides an attractive solution to the problem of rise of cell temperature [8].

Phase Change Materials

The basic superiority of the PCM is due to its excellent heat storage capacity per unit mass at a constant temperature hence the latent heat of the selected PCM should be as high as possible. High thermal conductivity is required for efficient conduction of heat through the body of PCM. Small volume changes on phase transformation and small vapor pressure at operating temperatures are desirable to reduce the containment problem. High density PCM would require a smaller size of storage container. Solid-liquid PCMs involve a relatively smaller change (10% or less) in volume and do not have any technical limitations [9-11]. A large number of such PCMs with different temperature ranges are commercially available.

The selected PCM must be non-flammable or the flammability must be very low. They must be non-toxic, non-corrosive and have good thermal stability as measured in terms of the thermal cycles. The economic constraints [10, 12-14] are price, recycle-ability and abundance. These factors influence the choice of PCM.

Classification of Phase Change Materials

The solid-liquid PCMs are classified into three main categories; organic, inorganic and eutectic [8-9] as shown in Figure 2.

Out of all these categories, the most suitable material is organic paraffins, which are available in a wide range of temperatures along with stable, chemical and physical properties. They are non-toxic, safe, generally non-corrosive and non-flammable and available abundantly at reasonable cost [9-11]. Also, their volumetric change while undergoing phase change is small and they possess a low vapor pressure when melted.

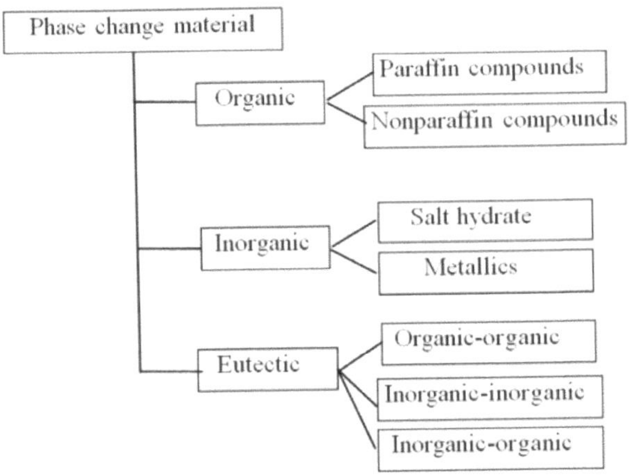

Fig. 2: Classification of PCMs [8-9]

Table 2: Thermo Physical Properties of some Paraffin Organic PCMs in the Range of 25 to 40°C

S. No.	*Material*	*Melting/ Freezing Point (°C)*	*Latent Heat (kJ/kg)*	*Density Liquid/ Solid (kg/m³)*	*Thermal cond. Liquid/ Solid (W/mK)*	*Flammability*	*Thermal Stability (Cycles)*
1.	OM29	29/26	194	870/976	0.17/0.29	Y	~2000
2.	OM30	32/30	230	878/906	0.12/0.18	N	~2000
3.	OM32	33/30	157	870/926	0.14/0.21	N	~2000
4.	OM35	35/34	171	870/900	0.16/0.20	N	~2000
5.	OM37	37/36	179	860/973	0.13/0.16	N	~2000

Table 3: The Hourly Variation of Temperature at Jodhpur on May 28, 2021 [16]

Time of Day (hours)	*Temperature (°C)*	*Time of Day (hours)*	*Temperature (°C)*
0.30	32.2	12.30	37.8
1.30	32.2	13.30	40.0
2.30	**31.1**	14.30	41.1
3.30	**31.1**	15.30	41.1
4.30	**31.1**	16.30	**42.2**
5.30	**30.0**	17.30	**42.2**
6.30	**30.0**	18.30	40.0
7.30	**31.1**	19.30	37.8
8.30	32.2	20.30	37.2
9.30	32.8	21.30	36.1
10.30	35.0	22.30	35.0
11.30	36.1	23.30	33.9

It can be concluded that for cooling of PV panels around Jodhpur, the organic PCMs; particularly paraffin should be preferred owing to appropriate desirable characteristics. Considering these factors and temperature range of 25–40°C in application of interest, the possible PCMs are listed in Table 2 [15].

Selection of PCM

As stated earlier, Jodhpur experiences very high daytime temperature in the months of summer. Table 3 gives the hourly variation of temperature at Jodhpur on May 28, 2021, the warmest day observed in summer 2021 [16]; notable in the table are the maximum and minimum temperatures, which control the selection of PCM as discussed below. The same is plotted versus the time of the day in Figure 3. Since a significant part (1-efficiency of the PV cell ≈0.80–0.85) of incident solar radiation on the panel surface is converted into heat, the temperature of the PV panel is higher than the ambient temperature during the day.

Two horizontal lines corresponding to temperatures 32°C and 36°C are drawn in Figure 3, which controls the decision of the selection of PCM to be used with reference to the ambient temperature considered here. Corresponding to the two extremes depicted by the horizontal lines, we discuss the use of PCM1 and PCM2 with melting temperatures as 32°C and 36°C, respectively.

As far as heat removal from the PV cell is concerned, the melting temperatures of both PCM1 and PCM2 are significantly lower than ambient temperature during the daytime when the PV cell is generating electrical power; the PCM1 would be better because of a larger temperature difference, as can be seen in Figure 3.

During the off-sun hours (evening to morning around 9 AM) when the PV panel and the PCM used for heat removal reject heat to the surroundings, the PCM must solidify. It is to note that the major portion of the heat of the panel is removed by the PCM by its melting and solidification processes and a small fraction of the panel heat may be removed by the sensible heating and cooling of the PCM. Here two important factors, which control the solidification process, are to be considered:

1. The duration available for cooling (the period during which the ambient temperature would be lower than the solidification temperature of the PCM), and
2. The difference between the solidification and ambient temperatures (for illustration marked in the figure at one place as ΔT for PCM2), which controls the rate of heat transfer and hence the solidification process. The heat transfer rate is proportional to ΔT.

It can be readily seen that for PCM1 the ambient temperature is lower than the solidification temperature during 2–8 AM, i.e., for about 6 hours and the average temperature difference ΔT during this period is only 1.5°C, which is significantly low for heat transfer and the solidification of the PCM1 may not take place; the corresponding values of duration and temperature difference for the PCM2 are 22.30–10.30, i.e., 12 hours and average temperature difference for heat rejection $\Delta T = 4$°C.

In Figure 3, the likely variation of the melting, solidification and sensible cooling processes of PCM2 with time are shown by a dashed line. Thus, any PCM selected must have melting temperature around 36°C; it is to note that lower the melting temperature of the selected PCM (provided the solidification during cooling period happens), lower will be the working temperature of the solar panel and better its efficiency would be.

The melting and cooling processes have been discussed here with respect to the ambient temperature. As mentioned earlier, during the daytime the panel temperature will be higher than the ambient temperature and thus the heat transfer from the heated panel to the PCM will be enhanced. But during the cooling process of the panel, the panel temperature will be higher than the ambient temperature for some duration and then at certain point of time, its temperature lowers to an extent that will depend on:

1. The ambient temperature and wind velocity for cooling by convection heat transfer process, and
2. The sky temperature (of the order of –10°C to –20°C at Jodhpur), which controls the radiation heat loss from the panel surface and hence the temperature of the panel during the cooling.

Thus, this process becomes a complex problem of heat transfer, which is further complicated by the low thermal conductivity of the PCM (0.13–0.16 W/mK) that controls the conduction heat flow through the PCM material.

To increase thermal conductivity of the PCM, dispersing high thermal conductivity particles, such as carbon, metals, or graphite within PCMs and impregnation of PCM in expanded graphite has been considered [17]. So, if the PCM with enhanced thermal conductivity is used, the cooling of the panel would be quite effective. and their effect on the performance of the PV panel would be considerable.

Looking to the complexity involved in the heat transfer from and to the panel and PCM, and the recommended use of high thermal conductivity particles dispersing within the PCM, an experimental study has been planned and is presently in the process of development.

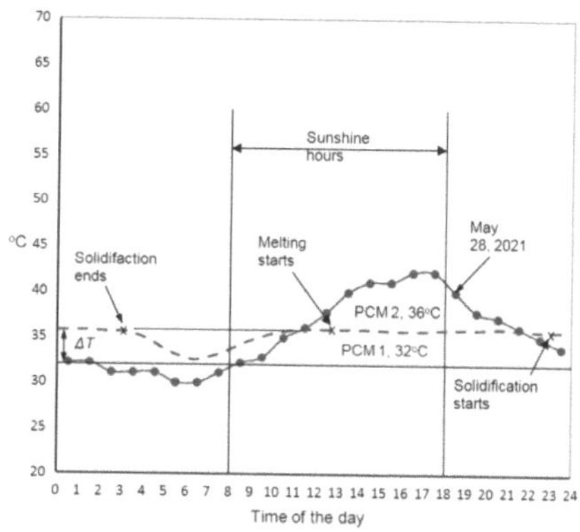

Fig. 3: The Ambient and PCM Temperatures versus Time of the Day; Showing Melting, Solidification and Cooling of PCM in Ambient

The study would focus on the pattern of melting, solidification and sensible cooling of the PCM with and without thermal conductivity enhancement.

Conclusion

The effect of working temperature of the PV cell on its output and the need of the effective cooling in the summer conditions of Jodhpur have been discussed in this paper, followed by a brief introduction to the characteristics of various active and passive cooling strategies reported in the literature. Looking to the superiority of PCM in cooling, a detailed discussion on the controlling factors for the selection of PCM with reference to the extreme ambient summer temperatures have been presented. A systematic approach for the selection of appropriate PCM for the climatic conditions of Jodhpur has been demonstrated by giving special consideration to summer season.

References

[1] Progress of solar energy in Rajasthan [date of access 25.10.21] https://energy.rajasthan.gov.in/content/raj/energydepartment/rrecl/en/projects/completed.html

[2] Climate of Rajasthan [date of access 20.10.21] https://imdpune.gov.in/library/public/Climate%20of%20Rajashtan.pdf

[3] Skoplaki, E. and Palyvos, J.A., on the temperature dependence of photovoltaic module electrical performance: A review of efficiency/power correlations, *Solar Energy,* Vol. 83, No. 5, pp. 614–624, 2009. https://doi.org/10.1016/j.solener.2008.10.008

[4] Masters, G.M., *Renewable and Efficient Electric Power Systems*, John Wiley & Sons, Inc., Hoboken, New Jersey, 2004.

[5] Khyani, H.K., Gehlot, V. and Joshi, B., Efficiency enhancement of photovoltaic modules: A review of cooling techniques, Conference Proceedings of National Conference on Recent Advances in Environmental Sciences and Engineering Technologies (RAESET-2016), pp. 137–141, Nov. 2016. http://raeset.jietjodhpur.ac.in/confere-nceproceeding.pdf

[6] Trina Solar Polycrystalline Panel TSM 290PC/PA14 https://makeskyblueusa.com/wp-content/uploads/2020/07/Trina-TSM-PA14_US_285-305_Silver.pdf

[7] Dwivedi, P., Sudhakar, K., Soni, A., Solomin, E. and Kirpichnikova, I., Advanced cooling techniques of PV modules: A state of art, *Case Studies in Thermal Engineering*, Vol. 21, 100674, 2020. https://doi.org/10.1016/j.csite.2020.100674

[8] Sharma, A., Tyagi, V.V., Chen, C.R. and Buddhi, D., Review on thermal energy storage with phase change materials and applications, *Renewable and Sustainable Energy Reviews*, Vol. 13, No. 2, pp. 318–345, 2009. https://doi.org/10.1016/j.rser.2007.10.005

[9] Zhou, D., Zhao, C.Y. and Tian, Y., Review on thermal energy storage with phase change materials (PCMs) in building applications, *Applied Energy,* Vol. 92, pp. 593–605, 2012.

[10] Akeiber, H., *et al.*, A review on phase change material (PCM) for sustainable passive cooling in building envelopes, *Renewable and Sustainable Energy Reviews*, Vol. 60, pp. 1470–1497, 2016. https://doi.org/10.1016/j.rser.2016.03.036

[11] Vakhshouri, A.R., Paraffin as phase change material, *Intech Open Book Series*, December 2019.

https://www.intechopen.com/chapters/70480

[12] Lane, G.A., *Solar heat storage: latent heat Materials. Background and principles*, Vol. 1. Boca Raton, FL: CRC Press Inc.; 1983.

[13] Konuklu, Y., Şahan, N. and Paksoy, H., *Latent Heat Storage Systems*, Editor(s): Ibrahim Dincer, Comprehensive Energy Systems, Elsevier, pp. 396–434, ISBN 9780128149256, 2018.https://doi.org/10.1016/B978-0-12-809597-3.00231-5

[14] Farid, M.M., Khudhair, A.M., Razack, S.A.K. and Hallaj, S.A., A review on phase change energy storage: materials and applications, *Energy Conversion and Management,* Vol. 45, No. 9:10, pp. 1597 1615, 2004. https://doi.org/10.1016/j.enconman.2003.09.015

[15] PCMs Product Range [date of access 20.10.21] https://www.pluss.co.in/product-range-PCM.php

[16] Weather in Jodhpur, May 2021 [date of access 20.10.21]
https://world-weather.info/forecast/india/jodhpur/28-may

[17] Min Li, A nano-graphite/paraffin phase change material with high thermal conductivity, *Applied Energy*, Vol. 106, pp. 25–30, 2013.

Influence of Time Delays on Performance Measures of Bioreactors

Kavirayani Srikanth
Department of Electrical and Electronics Engineering,
Gayatri Vidya Parishad College of Engineering (A),
Madhurawada Visakhapatnam, Andhra Pradesh–530048, India
E-mail: kavirayanisrikanth@gmail.com

ABSTRACT: *This paper investigates the bioreactor performance under the influence of time delays. The controllers designed would work efficiently without any delay in inputs but the presence of unwanted time delays that enter the system might as well influence the performance of the bioreactor which is investigated as part of this work. It is shown that the controller can still be stabilized without much performance hindrance under the influence of time delays with suitable feedback controllers.*

Keywords: Bioreactor; Controller; Feedback.

Introduction

The usage of bioreactor plants includes industries ranging from food manufacturing to pharmaceuticals industries and not limited to beverage manufacturing units. Normally, there are various classes of controllers that are used with these bioreactor plants. The stability of the plants is a major concern in the 24×7 operation of this manufacturing units which are fully automated. Any process delay would as such be a large problem as it might cause interrupts.

Sundari *et al.*[1] has taken the plant model of a unstable bioreactor process and have used classical control techniques to stabilize the plant and have considered various operating conditions such as biomass concentration, substrate concentration for the process control design using standard LQR and have analyzed the dynamic behavior of the reactor. The observable state estimation of immeasurable uncertain bioreactors were analyzed by Zhang *et al.*[2]. Typical assumptions made in nonlinear time delay dynamical system analysis of process inside and outside bioreactor are made with certain assumptions such as no medium is pumped inside and outside of a bioreactor by yuan *et al.*[3] for robust optimization process design.

This paper investigates the design of a unstable bioreactor process with system parameters designed by hernan *et al.*[4] which is taken with some assumptions. However, in view of the analysis in terms of time delay effects this paper elaborates on the influence of time delay with state matrices redefined for inclusion of time delay which thereby influence the performance of the unstable plant. The plant parameters are however not varied and are assumed to be constant for the ease of understanding. Authors [6-8] have worked on different aspects of bioreactors ranging from design to comparison between different reactor types and also have indicated the recent challenges with the reactors and applications. This work looks at a different perspective looking it from a mathematical design view which brings in new insights into the development of the system studies which will help us study the influence of delays with a more serious approach

Methodology

1. ***Bioreactor Plant Model***: The unstable bioreactor plant space model considered is as follows as taken from: [1]

$$\dot{X} = AX + Bu \quad \dots (1)$$

$$Y = CX + Du \quad \dots (2)$$

where

$$A = \begin{bmatrix} \mu - D_s & x_{1s}\,\mu_s' \\ \dfrac{-\mu_s}{Y} & -Dx - \dfrac{\mu_s x_{1s}}{Y} \end{bmatrix}$$

$$B = \frac{x_{1S}}{x_{2f} - x_{2S}}$$

$$C = [0,1]$$

$$D = 0$$

The plant values for analysis are as follows for the state matrix, input matrix, output matrix and the transition matrix are as follows:

$$A = \begin{bmatrix} 0 & -0.0679 \\ -0.75 & -0.1302 \end{bmatrix}$$

$$B = \begin{bmatrix} -0.9951 \\ 2.4878 \end{bmatrix}$$

$$C = [0\ 1]$$

$$D = 0$$

The plant model is integrated with time delay which is varied for static constant time delay variations with a process of integration as defined by kavirayani[5].

2. ***Time Delay Effects***: The modified and updated state space model is obtained by considering the time delay model of the plant by row and column addition as defined by a technique defined in [5] which will integrate the time delay parameter into the system matrix. This results in the modified state transition formulation which will be given as followed in the equations defined in 3 and 4 which will update the model matrix into a new form which will enable it to analyze the system for better performance and understanding under intermittent delays which could be considered to be hindering the performance.

$$A\_New = \begin{bmatrix} 0 & -0.0679 & 0 \\ -0.75 & -0.1302 & 0 \\ 0 & 2/\tau & 4/\tau \end{bmatrix}$$

The corresponding matrices B and C are also modified to accommodate the matrix changes done in the state matrix correspondingly. τ is the time delay that is defined for analysis which is varied for different values to analyze the system performance which will define the time delay impact.

Results

Table 1 indicates below the variations of time delay that are tested on this plant model in Simulink environment of MATLAB R2013b Version which are tabulated here. The status Y indicates the recorded values that are present as part of this work with a Y for Yes and N for NO. The cases 4 to 6 have not been evaluated due to potential influence of performance on the performance in terms of computation times but might be considered with high end computing facilities.

The output performance of the case number 1 is given as follows.

Table 1: Time Delay Variations

Case. No.	*Appearance*		
	Time Delay Value (mS)	*Step Tested*	*Impulse Tested*
1.	0.002	Y	Y
2.	2	Y	Y
3.	20	Y	Y
4.	200	N	N
5.	2000	N	N
6.	20000	N	N

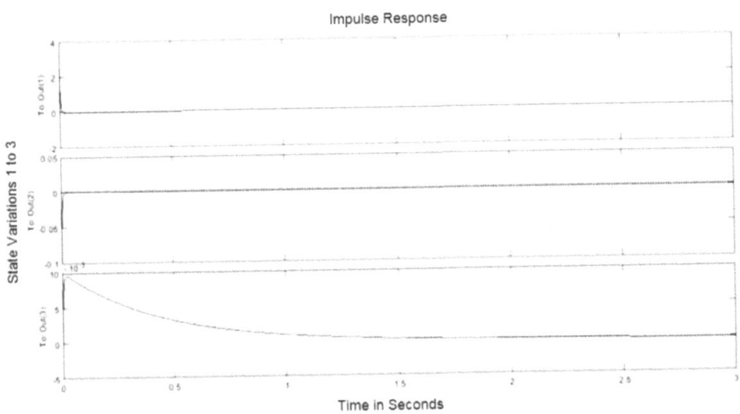

Fig. 1: State Variation for Impulse Response for 2 Seconds

Figure 1 indicates the variation of the new set of states defined with the time delay considered in the plant model as per the definition outlined in the matrix A_new. Figure 2 on similar lines indicate the variation in output for delay of 20 seconds as indicated in case iii.

Figure 3 on similar lines indicate the variation in output for delay of 0.002 seconds as indicated in case i.

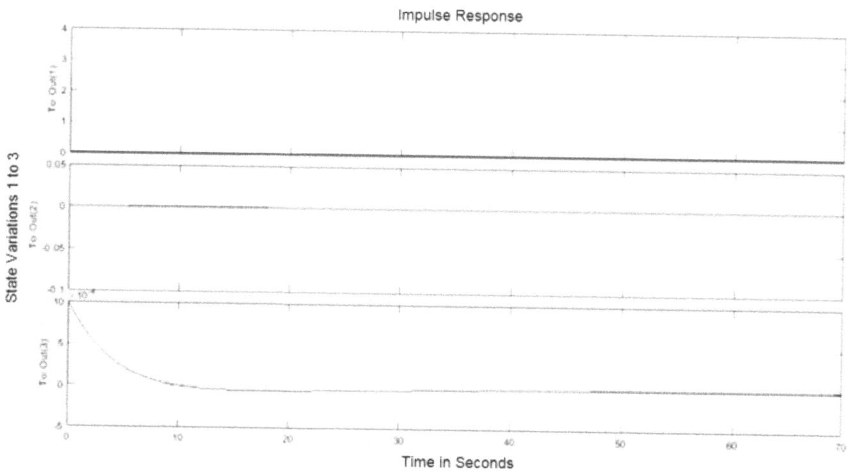

Fig. 2: State Variation for Impulse Response with a Delay of 20 Seconds

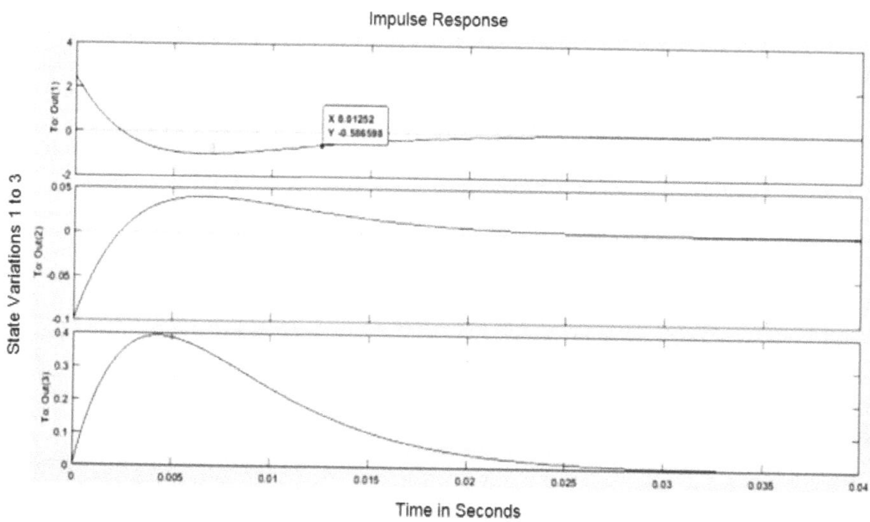

Fig. 3: State Variation for Impulse Response for 0.002 Seconds

Figure 4 is the nyquist plot of the system with integrated time delay parameters into the system definition which indicate the system to be stable irrespective of the time delay defined in the definition.

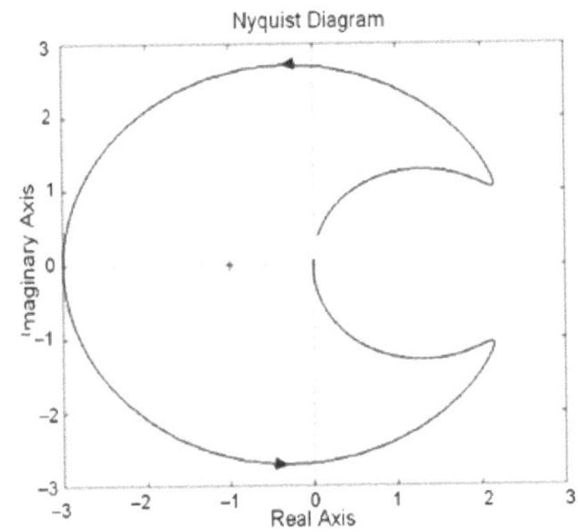

Fig. 4: Nyquist Plot with Integrated Delay

Conclusions

The conclusions that can be drawn from this work are that stability aspects of the bioreactor are highly sensitive. However, if time delay aspects are integrated into the system definitions and the analysis is extended using step and impulse response and with a standard controller such as LQR defined in the system the system performance would be normal to certain critical time limits which can be defined by the bifurcation laws that govern the stability parameters.

Acknowledgment

I thank the management of Gayatri Vidya Parishad College of Engineering (A) for the library facility and other technical computing facilities provided in the premises for the conduct of research.

References

[1] Sundari, S. and Neachiappan, Alamelu, "Modern Controller design for an unstable bioreactor process", Reference#59J-2018-04-03-01

[2] Zhang, J., Xia, Y., Shi, P. and Mahmoud, M.S., "New results on stability and stabilization of systems with interval time-varying delay," *IET Control Theory Appl.,* Vol. 5, No. 3, pp. 429–436, 2011.

[3] Yuan, Jinlong; Zhang, Xu; Liu, Chongyang; Chang, Liang; Xie, Jun; Feng, Enmin; Yin, Hongchao and Xiu, Zhilong," Robust optimization for nonlinear time-delay dynamical system of dha regulon with cost sensitivity constraint in batch culture, *Communications in Nonlinear Science and Numerical Simulation*, Vol. 38, 2016, pp. 140–171, ISSN 1007-5704, https://doi.org/10.1016/j.cnsns.2016.02.008. (https://www.sciencedirect.com/science/article/pii/S1007570416300272)

[4] Battista, Hernán De; Picó, Jesús; Garelli, Fabricio and Vignoni, Alejandro, Specific growth rate estimation in (fed-)batch bioreactors using second-order sliding observers, *Journal of Process Control*, Vol. 21, Issue 7, 2011, pp. 1049–1055, ISSN 0959-1524, https://doi.org/10.1016/j.jprocont.2011.05.008. https://www.sciencedirect.com/science/article/pii/S0959152411000989)

[5] Srikanth, Kavirayani, Control of underactuated systems with computational intelligence techniques under

the influence of time delays, Dissertation, GITAM {Deemed to be University}, 2019.

[6] Nadal-Rey, Gisela; McClure, Dale D.; Kavanagh, John M., Cassells, Benny; Cornelissen, Sjef; Fletcher, David F. and Gernaey, Krist V., Computational Fluid Dynamics modelling of hydrodynamics, mixing and oxygen transfer in industrial bioreactors with Newtonian broths, *Biochemical Engineering Journal*, 2021, 108265, ISSN 1369-703X, https://doi.org/10.1016/j.bej.2021.108265.

[7] Ahnen, Mathis von; Pedersen, Lars-Flemming and Pedersen, Per Bovbjerg, Formaldehyde degradation in denitrifying wood chip bioreactors: Effects of temperature, concentration and hydraulic retention time, Aquacultural Engineering, Vol. 95, 2021, 102199, ISSN 0144-8609, https://doi.org/10.1016/j.aquaeng.2021.102199.

[8] Bizon, Katarzyna and Tabiś, Bolesław, Problems in volumetric flow rate and liquid level control of a continuous stirred tank bioreactor with structured and unstructured kinetics, Chemical Engineering Research and Design, Vol. 175, 2021, pp. 309–319, ISSN 0263-8762, https://doi.org/10.1016/j.cherd.2021.09.015

Performance Improvement of MPPT based Grid Connected Hybrid Renewable Energy System

Bharat Bhushan Jain[1], **Rajkumar Kaushik**[2] **and Gori Shankar**[3]

[1,3]Department of Electrical Engineering, Jaipur Engineering College, Jaipur, Rajasthan–302028, India

[2]Department of Electrical Engineering, Amity University, Jaipur, Rajasthan–302028, India

E-mail: [2]rajkaushik1812@gmail.com

ABSTRACT: *The combination of renewable energy and electricity systems is growing. Solar photovoltaics and wind production has an important role in charging of batteries and grid applications over the past decade, because of many restrictions on natural sources of pollution energy inputs and environmental damage renewable energy sources need to be used. In India, oil is the primary and natural electricity supplier. However, the traditional energy crisis has forced the world to prepare for wind and solar systems. This work involves using wind and photovoltaic systems to design and extract the maximum power from DC networks under variable operating voltages. To increase the output power of the photovoltaic system and the wind generation system, it is necessary to find the highest possible energy from the photovoltaic system. In this work, the MPPT algorithm used for hybrid energy systems for maximum power perturbation and observation as well as the incremental conductance algorithm used for the hybrid system. P&O MPPT performance and incremental performance were compared and the results were analysed on MATLAB/SIMULINK platform.*

Keywords: MPPT, P&O, I&C, Solar and Wind Power System.

Introduction

For daily life and production, electricity is required. The power system comprises four sections: electricity generation, electricity transmission, power conversion and power consumption. In remote farms, forest farms and islands, transmission and transmission systems for small power demands are not easy to instal. Moreover, storms and other catastrophic events will continue to cause power outages [1]. Global warming is extremely serious and shifting renewable energy production is an excellent way to reduce fossil fuel emissions [2]. For these reasons, a renewable energy system must therefore be built outside the grid. One of the benefits of combining different sources of power is to provide sustainable power in areas not supplied by conventional grids [3]. They are essential in many applications, but hybrid energy systems are proposed to overcome this problem and make significant improvements because of their nonlinearity. The combination of multiple energy and storage units generally includes hybridization to optimize manufacturing and energy management [4]. Wind solar hybrid systems are used to generate electricity from renewable energies such as wind and solar energy. Solar panels and small generators of wind turbines are used for electricity generation in this installation [5].

To understand how a hybrid solar wind system works, we must first understand the functioning of a solar system and how it operates [6]. A solar system is a solar power system that produces electricity using solar panels. This figure shows a block diagram of a solar wind hybrid system for the generation of solar panels and wind turbines. Wind power is another source of renewable energy to generate electricity from wind turbines and generators.

A wind turbine is a two or three-blade fan rotating in the wind and aligned with wind direction [7]. A gearbox is a high-precision mechanical system used by mechanical methods to convert energy from one unit to another. Wind turbines of different shapes and sizes are available, but the horizontal and vertical axis turbines are most common [8]. Three main components are formed by solar energy systems: solar panels, photovoltaic solar cells and power storage batteries [9]. Solar panels can be used to store electrical energy batteries, to carry DC loads or to supply AC loads to the inverter. Only solar power is available during the day, while wind energy is always available depending on the weather [10].

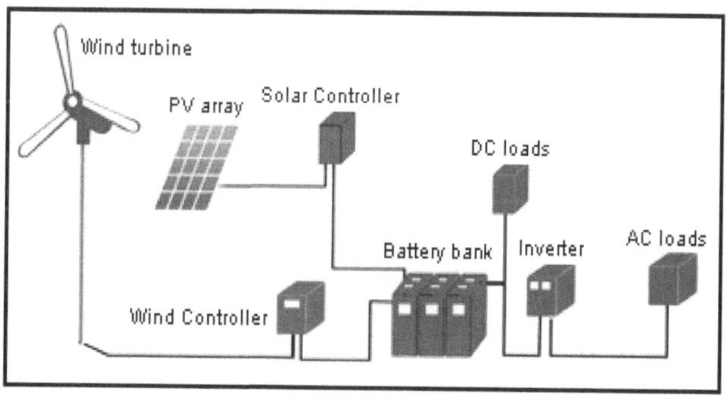

Fig. 1: A Small Size Solar-Wind Hybrid Power System [5]

Since wind and sun power complement each other, electricity can be generated almost throughout the year [11]. Wind Solar Hybrid system's main components include wind turbine and turbine generator, solar photovoltaic panels, batteries and wires [12]. The Wind-Solar Hybrid System produces electricity for battery charging and operating AC inverter [13]. The wind turbine is installed on a tower at least 18 m above ground level. Thanks to its height, the wind generator receives faster wind and greater power [14].

Objectives

The main objective of this study is to develop a MATLAB model for hybrid energy systems (solar, wind). This study aims to develop a single model of solar hybrid wind turbine, and then explores the feasibility, energy output and storage of the hybrid solar wind turbine system. Photovoltaic (PV) panels, wind turbines and energy storage batteries based on permanent synchronous generators are the components of the system. MPPT uses increased photovoltaic and wind power systems conductivity technology. The PV array is connected to a DC-DC converter amplifier after MPPT and connected toa standard bus network.

The objectives of conducting this thesis are:

- The objective of this work is to implement the grid-connected renewable energy system with Energy Storage System.
- To obtain the maximum power, perturb, observation, and incremental conductance-based MPPT algorithm applied to the hybrid system.
- To simulate the performance of the solar and wind hybrid system.

Maximum Power Point Tracking Algorithms

As the MPP of a solar panel varies with radiation and temperature, the MPPT algorithm is helpful in photovoltaic applications, so MPPT's algorithm is necessary to obtain the greatest energy needed for solar cells [15]. Many ways of finding MPPs have been discovered and published over the last decade. These technologies vary in many respects, such as sensor requirements, complexity, cost, efficiency of detection, speed, accurate tracking of changes in temperature and light, the required materials for application or personal information, etc. [16]. The other MPPT algorithms can be found in a complete overview [7]. These techniques include the most commonly used P&O and InCond algorithms.

Incremental Conductance

The incremental method of conductance is to measure and compare the terminal tension of the PV module with the direct driving. The maximum power is reached if the incremental conduction is equal to the instantaneous conduction. The terminal voltage of the PV module is regularly disturbed until the incremental conductance is equal to the immediate conductance. This can be seen in the next equation [17].

Algorithm for Incremental

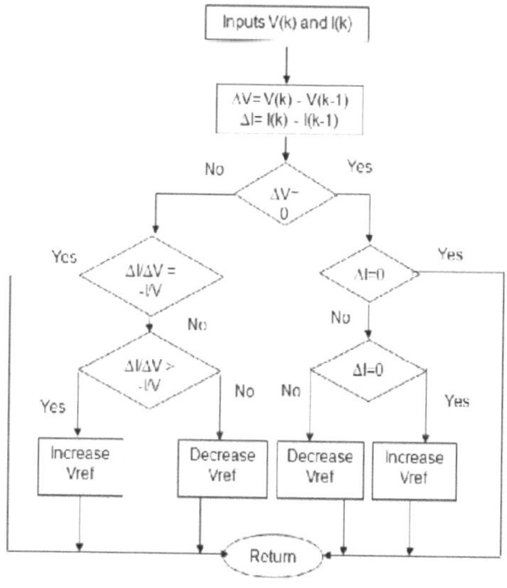

The incremental conductance method is explained in detail in chapter 4 for the maximum power point tracking for solar PV energy conversion system. The algorithmic steps are givenbelow.

Algorithmic Steps

Step 1: Sense the two consecutive voltages and current ofsolar PV.
Step 2: Calculate the dI/dV.
Step 3: If dI/dV > 0, the operating point is in the left of MPP. Increment the voltage.
Step 4: If dI/dV < 0, the operating point is in the right of MPP. Decrement the voltage.
Step 5: Go to step 1.

Perturb and Observation

The Observe and Perturb method for solar and wind power conversion systems is the most common one. In the solar photovoltaic photovoltaic output voltage and current, two consecutive intervals are measured. The power is calculated in succession for two intervals. The dP/dV voltage change is calculated. Based on positive and negative values of pitch dP/dV, the service cycle is increased or decreased. The voltage and power are therefore adjusted to the maximum power point. If the pitch dP/dV = 0, the maximum power point for current environmental conditions is reached. It's an ongoing process. The measures must be continuously measured and the change of power and voltage calculated to take control measures. The MPP is achieved by making the solar PV impedance with the load side impedance. The tariff cycle is adjusted according to the impedance. This MPPT algorithm is explained with a flow chart and algorithm in the previous chapter. Below are the algorithmic steps [18].

Algorithm of Perturb and Observation

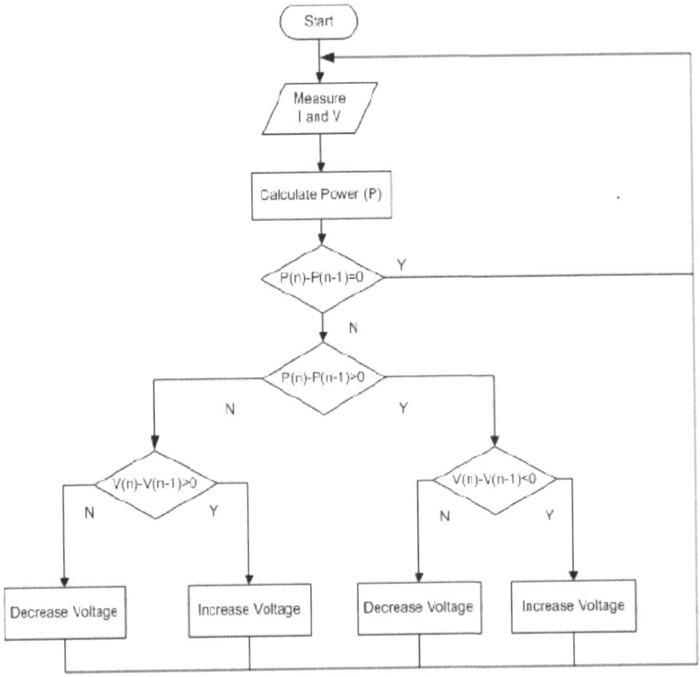

Algorithmic Steps

Step 1: Measure the two consecutive values of voltages and currents of solace PV.

Step 2: Calculate the powers $P(n)$ and $P(n–1)$.

Step 3: If the powers are increasing, then decreasethe duty cycle.

Step 4: If the powers are decreasing, then increasethe duty cycle.

Step 5: Go to step 1.

Methodology

The solar wind hybrid system components include a photovoltaic array, wind turbine, battery bench and inverter, controller and cable. Photovoltaic and wind turbines fulfill joint load requirements. If energy sources (solar wind) are abundant, solar energy continues to charge the battery until fully charged during the day. Unlike low energy, the battery generates energy that supports the photovoltaic range and the wind turbine until storage is depleted. The modeling of the solar-wind hybrid system is mainly dependent on the performance of each component. In order to forecast system performance, the two energy sources should be modeled separately and combined to meet demand reliability. When the power output from these sources is

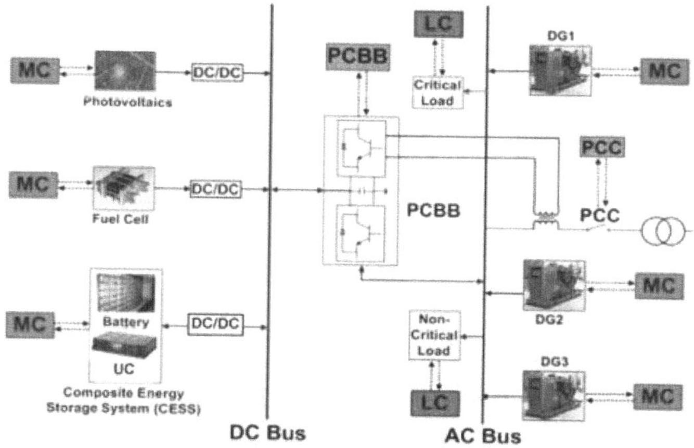

Fig. 2: Architecture Diagram

accurately predicted, the resulting combination provides efficient power. A hybrid system could be designed with electronic power interface to run in an isolated or network connected mode.

Simulation Results and Discussion

The MPPT algorithm is used in MATLAB Simulink for solar power and wind systems. MATLAB simulates wind and solar panels. Temperature and irradiance are taken as the input to the solar panel. The solar panel develops a temperature and irradiance-based voltage across its output terminals. The solar panel is connected to the boost converter. The converter matches the impedance of the solar panel to transfer maximum power from the solar panel to the load. The changeover boosts the load. When the load changes, the solar panel acts as a power source. MPPT shall be achieved through incremental conductance and disturbance and solar panel observationalmethods.

Table 1: Simulation Parameter

Simulation Parameters Value	*Values*
KP	40
Turbine Time Constant	0.05
Governor Time Constant	0.025
Electrical Generator (V.A.)	8.5E3/0.9
Engine Time Delay TD (S)	–0.024
Nominal Power	2E+006
Voltage	1
Frequency	50
Field Current	100
Open Circuit Voltage, VOC	0.62
Short Circuit Current	7.57
PV Output	230 KW/44I
Wind Power	4.5
Base Wind Speed (M/S)	12
Load Power (W)	100

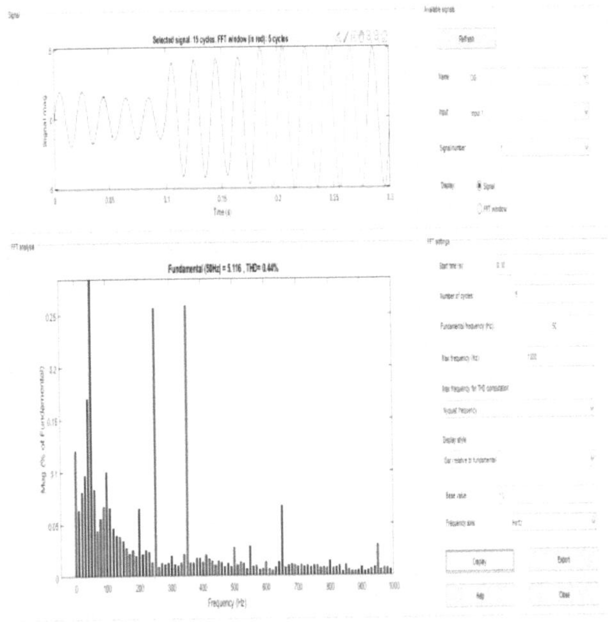

Fig. 3: THD Performance of Incremental and Conductance MPPT

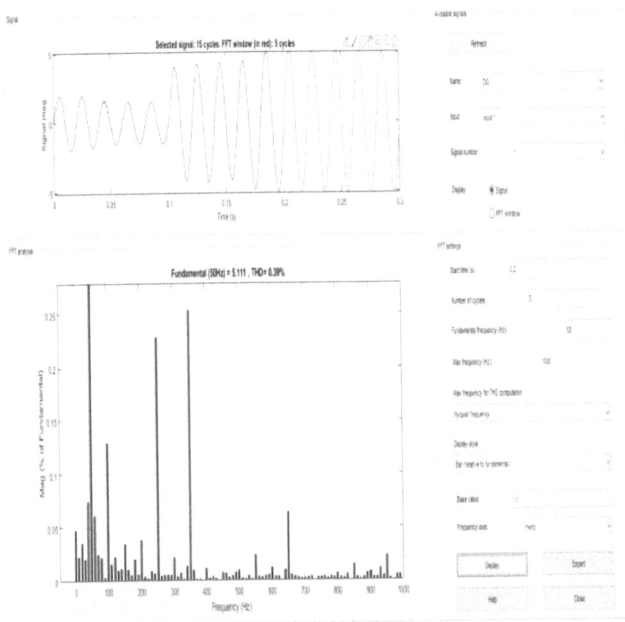

Fig. 4: THD Performance of Perturb and Observation MPPT

Table 2: Comparative Result Analysis

	MPPT	*THD*
Existing Technique	Fuzzy Logic Control	2.20%
Proposed Technique	Perturb and Observation	0.39%
	Incremental and Conductance	0.44%

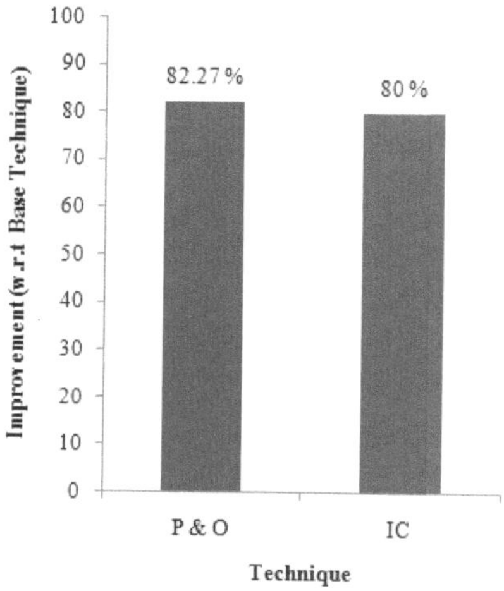

Fig. 5: Showing Percentage Improvement in THD w.r.t. Fuzzy Logic Control Scheme

Conclusion

The a alternative source of energy for conventional energy resources is renewable energy. The primary energy sources in remote areas where traditional energy cannot be transmitted are renewable energy sources. This thesis examines the hybrid energy system of the MPPT solar wind with a boost converter. A MPPT algorithm is used to increase the hybrid system's conversion efficiency. It is disturbed and observed algorithms and incremental conduct to provide the necessary tariff to manage the boost converter under unforeseen weather conditions. In this work, a detailed analysis of solar, wind and PMSG modelling was carried out. The simulation results indicate that dynamically and statistically the combination of a pitch angle controller, an inverter control and a grid-side inverter controller. The maximum power can be tracked and the wind turbine operated efficiently. The DC-link voltage is maintained in a stable level for active and reactive power disconnection. This optimizes grid output and consists of photovoltaic (PV) panels, wind turbines based on synchronous generators and batteries. MPPT uses increasing conductivity technology in photovoltaic and wind power systems. The PV array is connected to a DC-DC converter amplifier after MPPT and connected to a standard bus network.

The PV-Wind hybrid model is therefore proposed to compensate for the environmental and climate variability effects of the resources which impact continuous power generation. Single source renewable power generation can therefore meet load demands. For efficient tracking of solar and wind energy, Perturb and Observe (P&O) and IC MPP technology is used, and a boost converter is used to remove inverter fluctuations in the conversion of energy to AC. And a significant reduction in THD (82.2% and 80% using P&O and IC techniques) is achieved compared to existing control arrangements. The wind power system generates sinusoidal

AC power with a permanent magnet synchronous generator. The two energy sources are combined to power the grid to meet demands.

References

[1] Lasseter, R.H., "Microgrids," *IEEE Power Engineering Society Winter Meeting. Conf. Proc.*, Vol. 1, 2002, pp. 305–308.

[2] Anjali, R.K. Kaushik and Sharma, D., "Analyzing the Effect of Partial Shading on Performance of Grid Connected Solar PV System," *2018 3rd International Conference and Workshops on Recent Advances and Innovations in Engineering (ICRAIE),* 2018, pp. 1–4, doi: 10.1109/ICRAIE. 2018.8710395.

[3] Mohammed, A., Elgendy; Zahawi, Bashar and Atkinson, David J., 'Assessment of Perturb and Observe MPPT Algorithm Implementation Techniques for PV Pumping Applications,' *IEEE Transactions on sustainable energy*, Vol. 3, 2012, pp. 21–33.

[4] Kaushik, R.K., Mahela, O.P., Bhatt, P.K., Khan, B., Padmanaban, S.K. and Blaabjerg, F. "A Hybrid Algorithm for Recognition of Power Quality Disturbances," *in IEEE Access,* Vol. 8, pp. 229184–229200, 2020, doi: 10.1109/ACCESS.2020. 3046425.

[5] Yu, Ting-Chung and Lin, Yu-Cheng, 'A study on Maximum power point tracking algorithms for Photovoltaic Systems', *International Conference. Trishan Esram & Patrick L Chapman 2007,* 'Comparison of Photovoltaic Array Maximum Power Point Tracking Techniques', *IEEE Transactions on Energy Conversion,* Vol. 22, 2012, pp. 439–449.

[6] Babayeva, M., Abdullin, A., Polyakov, N. and Konstantin, G., "Comparative Analysis of Modeling of a Magnus Effect-Based Wind Energy System," 2020 *IEEE Conference of Russian Young Researchers in Electrical and Electronic Engineering (EIConRus),* 2020, pp. 602–605, doi: 10.1109/ EIConRus49466.2020.9039233.

[7] Kaushik, R.K., Mahela, O.P., Bhatt, P.K., Khan, B., Gag, A.R., Alhelou, H.H. and Siano, P. "Recognition of Islanding and Operational Events in Power System with Renewable Energy Penetration Using a Stockwell Transform-Based Method," *in IEEE Systems Journal,* doi: 10. 1109/JSYST.2020.3020919.

[8] Xing, C., Xi, X., He, X. and Liu, M., "Research on the MPPT Control Simulation of Wind and Photovoltaic Complementary Power Generation System," *2020 IEEE Sustainable Power and Energy Conference (iSPEC),* 2020, pp. 1058–1063, doi: 10.1109/iSPEC50848.2020.935 0965.

[9] Ji, G. and Ohyama, K., "MPPT control of Variable Speed Wind Power Generation System using Switched Reluctance Generator and AC-AC converter," 2020, *23rd International Conference on Electrical Machines and Systems (ICEMS),* 2020, pp. 1012–1016, doi: 10.23919/ ICEMS50442.2020.9290911.

[10] Bedoud, K., Merabet, H., Bahi, T. and Drici, D., "Fuzzy Observer for MPPT Control of Variable Wind Energy Conversion System Associated to AC-DC Converters," 2020 *20th International Conference on Sciences and Techniques of Automatic Control and Computer*

Engineering (STA), 2020, pp. 231–236, doi: 10.1109/STA50679.2020.9329352.

[11] Mishra, J., Pattnaik, M. and Samanta, S., "Power Management Scheme for a Wind-Photovoltaic Hybrid Autonomous System with Battery Storage," 2018 *IEEE 4th Southern Power Electronics Conference (SPEC),* 2018, pp. 1–5, doi: 10.1109/SPEC.2018.8635962.

[12] Kaushik, R.K., Mahela, O.P. and Bhatt, P.K. "Hybrid Algorithm for Detection of Events and Power Quality Disturbances Associated with Distribution Network in the Presence of Wind Energy," *2021 International Conference on Advance Computing and Innovative Technologies in Engineering (ICACITE),* 2021, pp. 415–420, doi: 10.1109/ICACITE51222.2021.9404665.

[13] Jagwani, S., Sah, G.K. and Venkatesha, L., "MPPT Based Switched Reluctance Generator Control for a Grid Interactive Wind Energy System," *2018 7th International Conference on Renewable Energy Research and Applications (ICRERA),* 2018, pp. 998–1003, doi: 10.1109/ICRERA.2018.8566748.

[14] Kaushik, R.K., Singh, A.K. and Sharma, J., "Study & Analysis of Incremental Conductance Method of Maximum Power Point Tracking System", Published in *International Journal of Engineering Research and Generic Science (IJERGS),* Vol. 3, Issue 1, January–February 2017, pp. 26–33, ISSN 2091-2730.

[15] Fu, C., Pan, T., Liu, H., Wu, D., Shen, Y. and Hao, Z., "MPPT Control Based Fuzzy for Wind Energy Generating System," *2018 37th Chinese Control Conference (CCC),* 2018, pp. 7465–7470, doi: 10.23919/ChiCC.2018.8484096.

[16] Kumar, D., Kaushik, R.K. and Sharma, D. "Analysis and Implementation of Phase Disposition Pulse Width Modulation Cascaded Hybrid Bridge 15 level Inverter" published in *International Journal of Engineering Science Technology and Research (IJESTR),* Vol. 4, Issue 4, July–August 2019, pp. 15–24, ISSN 2456-0464.

[17] Saravia, N.R.B., Gamarra, J.H.G., Infantes, J.A.A., Jimenez, S.L.R. and Zamata, J.R.M., "MPPT Design and Control of a Small Power Autonomous Photovoltaic Solar System: Application to a 90 W Two-Axis Autonomous Solar Photovoltaic Module," *2019 IEEE International Autumn Meeting on Power, Electronics and Computing (ROPEC),* 2019, pp. 1–6, doi: 10.1109/ROPEC48299.2019.9057069.

[18] Kumar, D., Kaushik, R.K. and Sharma, D. "A Novel Modeling Approach of Phase Disposition Pulse Width Modulation Cascaded Hybrid Bridge 15 level Inverter Using MATLAB Simulink" published in *International Journal of Innovative Engineering and Emerging Technology (IJIEET),* Vol. 5, Issue 4, July–August 2019, pp. 01–08, ISSN: 2455-3182.

Particle Swarm Optimization (PSO) based Design of Optimal PID Controller for Higher Order Systems

Khamma Kanwar[1], Jayashri Vajpai[2] and Santosh Kumari Meena[3]

Department of Electrical Engineering, M.B.M. Engineering College,
Jai Narain Vyas University, Jodhpur–342011, Rajasthan, India
E-mail: [1]kkjoya.jnvu@gmail.com; [2]jayashrivajpai@gmail.com;
[3]sanumeena11245@gmail.com

ABSTRACT: *Particle Swarm Optimization (PSO) is an emerging non gradient optimization technique based on swarm intelligence. In this paper, optimal PID controller design methodology that makes use of PSO has been developed and implemented for the higher order systems. MATLAB/Simulink software has been used to implement the proposed methodology for the design of PID controllers for higher order systems represented by their transfer functions and their Step Response Characteristics are compared with those obtained by use of the classical Ziegler-Nichols (Z-N) method based PID controller. The result obtained from the implementation of both methods is compared in terms of tuning parameters of PID controller, peak overshoot (%), and settling time. It has been found that the PSO based controller has better performance than Z-N based controller.*

Keywords: Particle Swarm Optimization (PSO), Ziegler-Nichols (Z-N) Method, PID Controller, MATLAB/Simulink.

Introduction

Proportional Integral Derivative (PID) control is the most common form of feedback control. Application of PID controllers can be found in all areas where modern control tools are used from early governors.

Nowadays several innovative techniques are available for the design of PID controller. In addition to the classical methods, these include soft computing methods such as Fuzzy logic, Neural Networks, Genetic Algorithm, Ant Colony Optimization, Particle Swarm Optimization (PSO).

In spite of the simple structure and robustness of PID controller, optimal tuning of gains K_p, K_I, K_d have been quite difficult. M. Nasri *et al.* [1] have presented a PSO method for determining the optimal PID controller parameters, for speed control of a linear brushless DC motor. This approach has superior features like easy implementation, stable convergence characteristics, and good computational efficiency. In comparison with the Genetic algorithm (GA) and Linear quadratic regulator (LQR) method, the proposed method was more efficient in improveing the step response characteristics such as reducing the steady-state error; rise time, settling time, and maximum overshoot in speed control of a linear brushless DC motor.

X. Lei *et al.* [2] have solved an economic multi-objective optimization problem through two improved intelligent optimization methods, one is multi-objective genetic algorithms named RNPGA (Random Niche Pareto Genetic Algorithms), and the other is improved PSO with linear inertia weight reduction. The results show that the PSO algorithm is faster to solve the problem.

In 2012 V. Rajinikanth and K. Latha [3] have designed an optimal controller for unstable systems using Evolutionary Algorithm. The simulation results show that the PSO based PID controller tuning approach provides improved performance for the setpoint tracking, load disturbance rejection, error minimization, and measurement noise attenuation for a class of unstable systems.

Inspired by these, an attempt has been made to develop a PSO based methodology for the optimal design of PID controller for higher order systems, as the design of controllers for such systems by using the conventional techniques is very tedious and often leads to significant error in performance.

In this paper, a methodology for PSO based optimal design of PID controller is proposed for different types of higher order plants represented by their transfer functions.

Design of PID Controller

The PID controller calculates "error" values as the difference between a measured process variable and the desired set point. It attempts to minimize the error by adjusting the process control inputs. The design of the PID controller involves three separate constant parameters and is accordingly called three term control: the proportional, the integral and derivatives values, denoted as P, I and D respectively [2].

The transfer function $G_c(s)$ of the PID controller can be described as:

$$G_c(s) = K_p + \frac{K_i}{s} + K_d s \qquad \dots (1)$$

By tuning the three parameters in the PID controller algorithm, the controller can provide control action designed for specific process requirements.

Ziegler-Nichols Method

The Z-N method can be used for both closed and open loop systems.

Z-N Open Loop Method (Process Reaction Curve Method)

This tuning formula for PID controller is based on experimental step responses or on the value of K_p that results in marginal stability provided only Proportional control is in action.

Experimental response of the plant to a unit step input is obtained and analyzed. If the plant involves neither an integrator nor dominant complex-conjugate poles, then such a unit-step response curve looks like '*S*' in shape. The S-shaped curve may be characterized by two constants, delay time (L) and time constant (T). *Z-N* suggested the timing of setting the values of K_p, T_i, and T_d according to the formula described in Table 1 [4].

Table 1: Z-N Tuning Rule Based on Step Response of Plant [4]

Type of Controller	K_p	T_i	T_d
P	$\frac{T}{L}$	∞	0
PI	$0.9\frac{T}{L}$	$\frac{L}{0.3}$	0
PID	$1.2\frac{T}{L}$	2L	0.5L

Z-N Closed-Loop Ultimate Gain Method

It is required to set $T_i = \infty$ and $T_d = 0$. Using the proportional control action, increase K_p from 0 to a critical value K_{cr} at which the first sustained oscillations an obtained in the output. If the output does not exhibit sustained oscillations for any value of K_p then this method is not applicable. Thus, the critical gain K_{cr} and the corresponding period P_{cr} in seconds are experimentally determined [4]. Z-N suggested the values of parameters K_p, T_i and T_d as depicted in Table 2 [4].

Table 2: Z-N Tuning Rule based on Critical Gain K_{cr} and Critical Period P_{cr} [4]

Type of Controller	K_p	T_i	T_d
P	0.5 K_{cr}	∞	0
PI	0.45 K_{cr}	$\frac{1}{1.2}P_{cr}$	0
PID	$0.6K_{cr}$	$0.5P_{cr}$	0.125P_{cr}

The formulae given in Tables 1 & 2 have been derived empirically for both the methods suggested by Ziegler and Nichols. They are quite successful in designing a controller that produces accurate responses for low order, linear, stable systems, but lose their relevance for higher order, nonlinear and unstable systems. Such systems require alternative methods for the design of controllers. These include intelligent and soft computing based techniques.

Particle Swarm Optimization

PSO is an emerging computational technique based on swarm intelligence. This method is motivated by the observation of social interaction behaviors such as fish schooling and bird flocking.

PSO was originally developed by Kennedy and Eberhart in 1995, taking inspiration both from the related field of evolutionary algorithms and artificial life methodologies. PSO learns from the problem solving scenario and uses it to direct the course of the optimization problems. In PSO every single solution is like a "bird" in the search space, which is called "Particle". And the population is called a swarm. All particles have fitness values which are evaluated by calculating the fitness function at the current location to be optimized and have velocities that direct the flying of the particles. The particles keep flying through the problem space by following the particles with the best solutions so far [5]-[7].

In the real number space, each individual possible solution can be modelled as a location of a particle that moves through the problem hyperspace. The position of each particle is determined by the vector $x_i \in R^n$ and its movement by the velocity of the particle $v_i \in R^n$ as shown in equation (2) & equation (3).

$$X_{id} = X_{id} + V_{id} \quad \dots (2)$$

The information available for each individual is based on its own experience (the decisions that it has made so far and the success of each decision) and the knowledge of the performance of other individuals in its neighborhood. Since the relative importance of these two factors can vary from one decision to another, it is reasonable to apply random weights to each factor and therefore the velocity will be determined by:

$$V_{id} = w * V_{id} + c_1 * rand_1() * (P_{id} - X_{id}) + c_2 * rand_2() * (P_{gd} - X_{gd}) \quad \dots (3)$$

Where

- w is the inertia weight factor.
- c_1 and c_2 are positive numbers, called Acceleration constant.
- $rand_1$ ( ) and $rand_2$ ( ) are two random numbers independently generated.
- p_i is the best position that the corresponding particle has found so far.
- p_g is the best position of the entire swarm [8].

Implementation of PSO based Design of Optimal PID Controller

The success of the PSO based design of optimal PID controller is highly dependent on the design of an effective fitness function which is instrumental in the proper selection of the gains of PID controller.

Fitness Function: The fitness function is designed for minimization of settling time and maximum overshoot as given in eq. (4). For this purpose, the step response of the system is calculated then it is used to calculate the error. Iterations are carried out till the error is minimized.

$$F = (1 - \exp(-0.5)) * (M_p + E_{ss}) + \exp(-0.5) * (t_s - t_r) \quad \dots (4)$$

Where

M_P = peak overshoot.

E_{ss}= steady state error.

t_s = settling time.

t_r = rise time.

PSO based Determination of PID Controller Gains

The PSO algorithm is used to find the optimal parameters for PID controllers. In the proposed PSO based method, each particle contains three members i.e. gains of P, I and D parts. It means that the search space has three dimensions and particles must 'fly' in a three dimensional space.

The following PSO parameters are used to design and tune the PSO based PID controller parameters:

Table 3: Parameter for PSO [1]

Parameter	*Values*
No. of Particles	20
No. of Iterations	20
Inertia weight factor (w)	0.6
Acceleration Constant (C_1)	1.5
Acceleration Constant (C_2)	1.5

The PSO algorithm is run with these parameters for the maximum iteration count of 20, so as to determine the optimal values of PID gains for the selected 3rd and 4th order plants as described in the next section.

Simulation Results

The closed loop PID controller cascaded with the system to be controlled was tuned for the optimal values of K_p, K_i, and K_d first by using the *Z-N* method and then by the PSO algorithm. Different types of plants represented by their transfer functions were explored as described in the following sections.

Plant I: 3rd order system [9]:

$$G(s) = \frac{1}{s^3 + 9s^2 + 23s + 15}$$

The values of the proportional, derivative and integral gains were derived for the plant by using Z-N method and then by PSO algorithm.

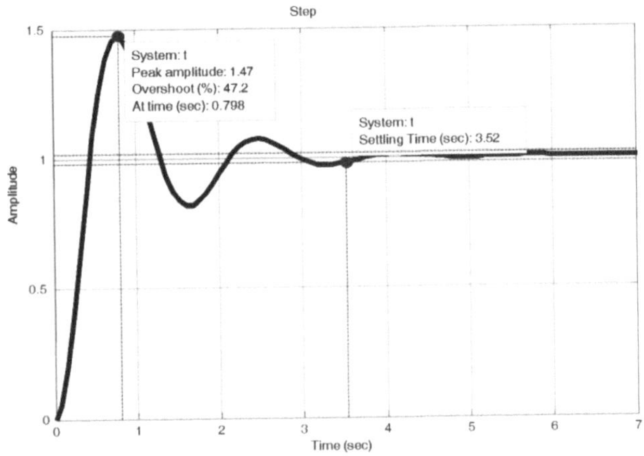

Fig. 1: Step Response of a Plant I with PID Controller Tuned by Z-N Method

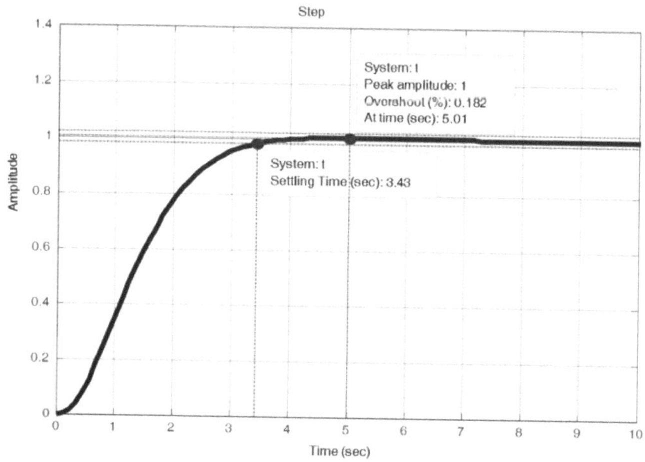

Fig. 2: Step Response of the Plant I with PID Controller Tuned by PSO Method

Plant II: 4th order system [10]:

$$G(s) = \frac{5}{s^4 + 3s^3 + 7s^2 + 5s}$$

The values of the proportional, derivative, and integral gains were derived for the plant by using the Z-N method and then by PSO algorithm.

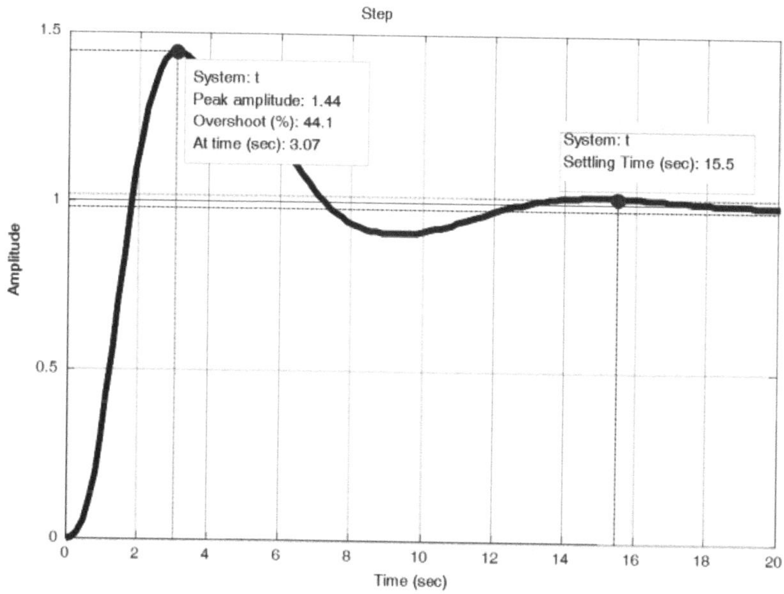

Fig. 3: Step Response of a Plant II with PID Controller Tuned by Z-N Method

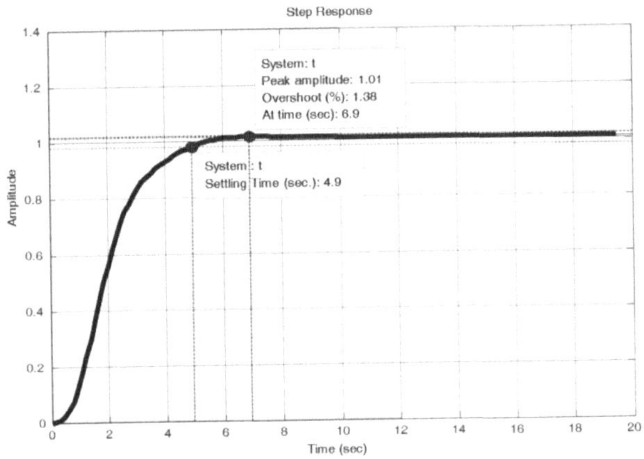

Fig. 4: Step Response of the Plant II with PID Controller Tuned by PSO Method

The PSO algorithm has been tested for tuning of PID controller for optimal settling time and peak overshoot of the plant transfer function and the results have been found as shown in Figures 1–4.

Table 4: Comparison of Settling time and Peak Overshoot Obtained by *Z-N* Method and PSO Algorithm

Method	*Plant I*	*Plant II*
Z-N Method		
K_p	115.2	0.9042
K_I	175.9	0.3603
K_D	18.9	0.6939
Settling Time (sec.)	3.52	15.46
Peak Overshoot (%)	47.2	44.1
PSO Algorithm		
K_p	10.8231	0.4905
K_I	10.4638	0.0015
K_D	0.1773	0.2982
Settling Time (sec.)	3.43	4.9
Peak Overshoot (%)	0.18	1.38

From the comparison of results shown in Table 4, it can be seen that by employing PSO based PID controller, the plant response produces lesser peak overshoot and settling time than the controller designed by using the classical Z-N method.

Conclusion

PSO based PID controller gives better transient response characteristics than any of the classical tuning methods. In this paper design of optimal PID controller for higher order systems were explored. The closed loop PID controller cascaded with the plant has been tuned for the values K_p, K_i and K_d first by using the Z-N method and then by PSO algorithm. Corresponding peak overshoot and settling time have been computed in both cases. It has been found that in all the cases, the peak overshoot and settling time obtained by PSO based controller is much less than the values obtained by the Z-N method based controller, as shown by comparison of result in Table 4.

References

[1] Nasri, M., Nezamabadi-pour, H. and Maghfoori, M., "A PSO-Based Optimum Design of PID Controller for a Linear Brushless DC Motor", *World Academy of Science, Engineering and Technology*, 2007.

[2] Lei, X., Shi, Z. and Wang, L., "Optimization and Simulation of a MO Problem Solved by GAs and PSO", *Third International Conference on Natural Computation (ICNC 2007),* 2007. DOI: 0-7695-2875-9/07.

[3] Rajinikanth, V. and Latha, K., "Tuning and Retuning of PID Controller for Unstable Systems Using Evolutionary Algorithm", *International Scholarly Research Network ISRN Chemical Engineering*, pp. 1–10, 2012. DOI: 10.5402/2012/693545.

[4] Ogata, K., "Modern Control Engineering", *Pearson Education Publisher*, 4th edition, 2002.

[5] Kennedy, J. and Eberhart, R., "Particle swarm optimization", *In Proceedings of IEEE International Conference on Neural Networks, Perth,* Australia, Vol. IV, pp. 1942–1948, 1995. DOI: 0-7803-2768-3/95.

[6] Shi, Y. and Eberhart, R., "A Modified Particle Swarm Optimizer", 1998. DOI: 0-7803-4869-9/98.

[7] Hu, X., Shi, Y. and Eberhart, R. "Recent Advances in Particle Swarm" 2004. DOI: 7803-8515-2/04.

[8] Eberhart, R. and Shi, Y., "Particle Swarm Optimization, Application, Resources" 2001. DOI: 7803-6657-3/01.

[9] Solihin, M.I., Tack, L.F. and Kean, M.L., "Tuning of PID Controller Using Particle Swarm Optimization (PSO)", *Proceeding of the International Conference on Advanced Science, Engineering and Information Technology,* ISBN 978-983-42366-4-9, 2011.

[10] Selvan, S.E., Subramanian, S. and Solomon, S.T., "Novel Technique for PID Tuning by Particle Swarm Optimization". https://www3.nd.edu/~swarm03/Program/Abstracts/SelvanSwarm2003.pdf

Generation Flexibility of Power System: A Review

Kumari Aishwarya, Prerna Jain and Abhishek Harit

Department of Electrical Engineering, Malaviya National Institute of Technology, Jawahar Lal Nehru Marg, Jhalana Gram, Malviya Nagar, Jaipur, Rajasthan–302017

E-mail: *2020psm5617@mnit.ac.in

ABSTRACT: *As India and other countries are working towards their low-emission goals, one of the most important resources to achieve this is using renewable energy. Large-scale integration of renewable energy makes system uncertain, and has variable generation. This uncertainty and variability in the power system requires a certain degree of flexibility. Power system flexibility is defined as the ability of a power system to use its own resources in order to respond to net load changes that are not met by variable generation. Renewable generations like wind, solar etc. vary at every second of time and it makes them inadequate to fulfill all the demand. Power system flexibility can be provided from different aspects like generation, transmission, and distribution. Ramping up/down and at lower level if the generators can work efficiently this directly impact the system flexibility. It includes generator characteristics as indicators such as operational range, ramp up/down, start up/down time, minimum up/down time etc. Flexibility in the generating units can be incorporated by either retrofitting the elements of the generating units or by optimizing the generator characteristics. The importance of generation flexibility, as well as the elements that influence generator flexibility can be seen.*

Keyword: Generation Flexibility, Renewable Energy, Fuzzy Analytical Hierarchy Process (FAHP), Unit Scheduling.

Introduction

Power system flexibility has become a global priority, with more and more integration of renewable generation in the power sector. Flexibility can be seen as a remedy for the rare abrupt changes and the rapid random fluctuation. Power system flexibility has many definitions as per different authorities like as IEA power system is flexible if it can be, within economic boundaries, respond quickly to high fluctuation demand and supply, ramping down when demand decreases and ramping up when increases [7]. However as the penetration of renewable energy increases the definition for the flexibility also changes as per the requirement It's a ability of power system to met the net load that are left or not met by the VRE that is Net load = total load – VR[1]. System said to be inflexible if there is a difficult in balancing demand and supply, significant renewable energy curtailment, area balance violation negative market price, price volatility etc. Requirements of the power system flexibility can be classified as flexibility for power, energy, voltage and transfer capacity [3].

- ***Flexibility for Power:*** It means frequency in the system should be maintain at standard level for short term period.
- ***Flexibility for Voltage:*** At predefined level voltage of bus should be maintained over short time period.
- ***Flexibility for Energy:*** Balance between demand and supply for demand scenario over medium to long term period.
- ***Flexibility for Transfer Capacity:*** Congestion in the system should be avoided by power transferability for short to medium time period.

For measuring flexibility present in the system many metrics has proposed till now some of them IRRE, Expected flexibility shortfall, Available flexibility distribution, period of flexibility deficit, expected unserved ramping [4],[10], [9],[17]. There are some composite metrics that also has been proposed using eight characteristics as a indicator [4].

Flexibility have different sources that can be seem form different perspective like from Generation side- generation flexibility, transmission side, distribution side, storage, sector coupling [5],[16]. Flexibility is more directly affected by generation flexibility if it ramp up or ramp down quickly. And taking the characteristics weight that influences the flexibility of the generators that also affect the system flexibility, flexibility also can be seen as the cost influencer in the unit scheduling [1].

Sources of Flexibility

Planning for flexibility necessitates taking into consideration all potential sources. Previously, power system networks achieved flexibility by deploying dispatchable power facilities (such as gas turbines) that could be turned on and off quickly. Flexibility is also achieved through various techniques as a result of the significant penetration of VRES. These techniques include demand side management (DSM), strengthening of transmission and distribution network, using storage systems, sector coupling and improved operations by load forecasting and scheduling [8].

Technical Flexibility

(a) *Generation Side*

Source-side flexibility is strongly linked to the performance of the technologies that make up a power system's generation fleet. The absolute power output range (MW), ramp rate (MW/min), and energy level continuity (MWh) are the three key characteristics that determine the flexibility of power systems. Before going for the other sources for flexibility we can see the effectiveness of flexibility of generators. Flexible generators have high ramping rates, quick start up and shut down and at the lower minimum level operation should be efficient. These properties are essential for generators when imbalance arises in the system, but using generators as a effective sources of flexibility the cost criterion important. Some generators like (hydro plants, gas-fired power plants, biogas, geothermal plants) have high flexibility with low cost [8].

(b) *Demand Response (DR)*

By demand response and demand management, load can be govern or incentives through pricing and use of the particular load. It can mitigate the imbalance of the grid or the system. Connected load can be cutoff when the price is high and or if there is a peak hour. Consumer participation in load control methods based on price signals would be encouraged and ensured when DSM and DR practices are implemented. Implementing DSM and DR has a need to set rules and norms in subject to legitimate demand-side resources minimum size, reaction time dependency. Smart grid, TOU tariffs, shifting of load, curtailed load, critical peak pricing, real-time pricing, smart metering, and centralized load management are some of the demand response activities [8],[2],[7].

(c) *Energy Storage*

Energy storage systems are used to store energy created by VRES during periods of excess and inexpensive electricity, and subsequently to use it when the demand arises. Peaking sources are typically storage devices. Energy storage can be done in large scale, medium or small scale. Large scale storage is thermal storage, pumped hydro storage and compressed air energy storage this large scale storage is done in network side and the medium and small scale storage is done in the distribution side. As a whole it increases the status of flexibility of system. Some other storage methods for distributed or decentralized scale are flywheel, electrolysis, super capacitors, fuel cell, batteries. Energy storage is very useful device for now a days that it can changes the timing for supply as it store the electricity. When the charges for the electricity is high it supply to the demand when its low it charges itself. This benefited the system by reducing the use of expensive generators and reduces the costs. It also facilitates the high share of VRE by mitigating the impacts of VRE on grid operations [5],[16].

(d) *Transmission Side*

Grid can be said as the aqueduct for generation and demand side flexibility as it net out the variance in the real time. It can be termed as a limiting factor if its flexibility is low. Ex-due to congestion problem, with high supply side flexibility, integrating high share of VRE generation may face difficulties. Grid flexibility has diverse meaning like robustness of transmission network that balance supply and demand in large area, secondly cross-border interconnection it enables exchange of flexibility across the border and also existence of advanced control of generators, advanced power flow control (FACTs), it enhances the communication among the elements [7],[16].

(e) *Sector Coupling*

Sector coupling provides the need for heat, electricity and ease of mobility. It includes power-to-gas (hydrogen, methane from electricity), power to heat (DHCs, heat pumps, resistors) and power to mobility (battery, electric vehicles) while assisting in the decarbonization of end-use sectors and flexibility. Example like coupling natural gas to electricity system. Electricity use natural gas as fuel when coal is not available. Natural gas use in gas load and also use in heat conversion by DHCs. This improve the energy transition and flexibility in the system [7],[16].

Operational Flexibility

Operation flexibility is termed as how the recourses in the system operates to provide the flexibility. It can be affected by the government policies, market strategy, system constraints and capability of resources, they are all inter related for effective working of systems [7].

It can be obtain through:

- Intraday market
- Co-optimize hydrothermal unit commitment
- Improved VRE forecasting
- Pooling of resources

In different time scale:

- In the long run, the system must be designed with enough operational flexibility to function well with a large amount of VRE. To make these viable, authorities may need to promote investment by employing capacity markets that reward flexibility, or by expanding the time granularity and space granularity of power markets, which would provide correct price signals to invest in long term.
- Long term to medium term scale includes seasonal variations and annual energy variations due to intermittent nature of climate.
- Medium to short term includes real time variation in generation, dispatching and scheduling of generating units. Designing of day ahead and intraday markets in critical order to optimize the system's flexibility.

Characteristics that Influence Flexibility

(a) *Ramp Up/Down:* Access rapid fluctuation in the power supply conventional generators requires rapid adjustment speeds in reply to the frequent changes in the net load. Ramp rate is the ability of generators to change its speed, from increasing ramp up or decreases ramp down rate in accordance with the minimum and maximum generation level. Ramping rate is directional and it also depends on the timescale and operating conditions and capacity of the units. Ramping conditions can be given as ramping up and ramping down as below equations, for example, a coal unit deployed at its minimal power output $P_{\min}$ will not be able to quickly modify its output to the maximum limit. Production limitation is also a governing term for maximum ramping limit apart from the dynamics of devices [5], [16].

$$P_{n.t} - P_{n,t-1} \leq UR_n$$

$$P_{n,t} - P_{n,t-1} \leq DR_n$$

where UR_n and DR_n are the ramping up and ramping down of the nth unit, respectively.

(b) *Generation Capacity:* Power output of thermal generators is in between the definite range and normal operating conditions. This range is between p_{min} and p_{max}. Running generators in it's p_{min} is normally not cost-elective, but it must operate at a minimal base load owing to the generator's minimum uptime limitation. Running in this range makes the generators to come to the net load variation take place in the system at any time. Generation capacity given as: [5].

$$P_{\min} < P < P_{\max}$$

(c) *Adjustable Capacity:* It is define d as the maximum adjustment that can be done by the thermal power unit (TPU) in the unit output to meet the net load at any time in the system under favor-able conditions. TPU's regulation capa-city is difference of the rated capacity and minimal stable generating level: [5].

$$S = P_{\max} - P_{\min}$$

(d) *Start Up /Down:* Unit sometime needs quick changes as supply or demand changes in the power system, recently this increases due to high penetration of renewable sources, however the thermal power plant have many constraints unlike the energy storage. Operating and maintenance costs of thermal power plant must be satisfied to do that generator must maintain scale of online operation. The typical TPU, on the other hand, is kept off for a period of time after the shutdown to avoid heat stress and other causes from diminishing the unit's life. The start-up, shut-down, and minimum up and down periods of a thermal unit are all directly connected to the adjustment period (T). T is the shortest time period between the time the TPU starts up and when it moves to $p_{\min}$, and it ends at the time the TPU starts moving to $p_{\min}$.[5],[16]

$$T = \min\left(T_{sut} + T_{mut} + T_{sdt} + T_{mdt}\right)$$

Sut – start up time
Mut – minimum up time
Sdt – start down time
Mdt – minimum down time

Role of Conventional Power Plant in Flexibility

Traditional power plants are the most common source of flexibility. Flexible power plant has a wide meaning of, fast responding power outputs, quick action of start and stop, reducing power out to minimum value without shutting down. Two different tactics that made generators more flexible are:[13],[6].

(a) *Changes to Operational Practices for Existing Plants:* To make the power plant more flexible, changes in the operational practices like data collection and real time monitoring can be done and it makes them free from more capital investment, this unlocks the flexibility in existing power plants.

(b) *Retrofitting Existing Plants:* To increase the different flexibility criteria of power plants, a variety of retrofit alternatives are available. For example, strategic coupling of BESS with existing facilities is rapidly becoming a feasible technological and economic way of enhancing flexibility ex natural gas peaking power plant 10 MW/4.3 MWh battery [6].

(c) *Some retrofitting options:* [6]

1. *Indirect Firing*—It lower minimum load, it together can decreases the minimum stable firing rate from 25–30%
2. Switching from two mill to single mill operation—Its lower minimum load, rump up increases Impact-resulted in reducing minimum load to 12.5%
3. Control system optimization – load 76% to 48%
4. New turbine start—Shorten start up time, its start up time can be reduced by 15% minimum
5. Thin walled component/special turbine design—shorten start up time, higher ramp rate.
6. Thermal energy storage for feed water preheating—Lower minimum load, impact—5–10% decrease.

FAHP Method

Min-up time, min-down time, ramp rates, minimum output of generation, operation range of generation, and restricted start-up periods are all examples of flexibility parameters in generators [5],[16]. The flexibility index for each generator is calculated using the FAHP approach in this article. The fuzzy analytical hierarchy process (FAHP) is used to determine the flexibility index of generating resources, and then the unit scheduling process is used to study the link between generation flexibility and unit scheduling cost under various scenarios [1],[5]. This method has different steps for the proposing of the metric firstly from selecting flexibility parameters, Normalization, weighting, correlation analysis. This is the framework for the calculation for the metric.

Calculation process given below.

A. *Normalization*

Factors that influences the weights of the parameters should be normalize. The min-max normalization process is chosen in this study as it converts all parameters into an identical range between 0 and 1 using Equation: [1], [5], [4].

$$I = \frac{x - \text{base}_{\min}}{\text{base}_{\max} - \text{base}_{\min}}$$

where I is normalization value, x is unit parameter, represents the value after normalization, $\text{base}_{\max}$ is the quantity that must be greater than or equal to the maximum value of all factors (reference value), and $\text{base}_{\min}$ is the quantity that must be less than or equal to the minimum of all factors.

B. *Weighting*

Weighting is done for indicators to assign the relative importance of the parameter in system flexibility. Weighting is classified as statistical or participatory that does depend on the data driven or expert opinion[1],[5].

C. *Consistency Index*

Consistency is a degree of inconsistency that is measured by quantitative indicator. Indicator (CI):[1], [5]

$$CR = \frac{\lambda_{\max} - 1}{n - 1}$$

$$n = \frac{N(N-1)}{2}$$

$\lambda_{\max}$ is largest eigen value of comparison judgment matrix. $CR < 0.1$ is random consistency ratio to correct the consistency index.

D. *Correlation*

If two indicators have a high degree of connection, attention should be given while assigning weights to them, as this may result in duplicate counting in the composite index. To address this problem, the mathematical correlation between each set of individual indicators x and y is first calculated using: [1],[5]

$$R_{xy} = \frac{\sum_i \left(x_i - \bar{x}\right)\left(y_i - \bar{y}\right)}{\left(n-1\right)\sigma_x \sigma_y}$$

where are n values of indicator x and y, $\bar{x}$, $\bar{y}$ are means and σ_x , σ_y are the standard deviations.

Unit Scheduling Considering Indicators

Renewable generation have variable generation taking that in the consideration for different scenario like it can be for different seasons or per day because now a days renewable is the upcoming sources for the generation and it is increasing day by day. It will give a basic or much to accurate relation between the flexibility and the need of it in the system, and maintaining that cost of the required flexibility, Shown in Figure 1 below [1],[5]. It makes the system reliabe and efficient when the flexibility is taken into consideration.

Table 1: Weighting Scale of FAHP [1]

Scales	*Definition*
0.5	Equally important
0.6	Slightly important
0.7	Obviously important
0.8	Very important
0.9	Extremely important
0.1–0.4	Relative importance

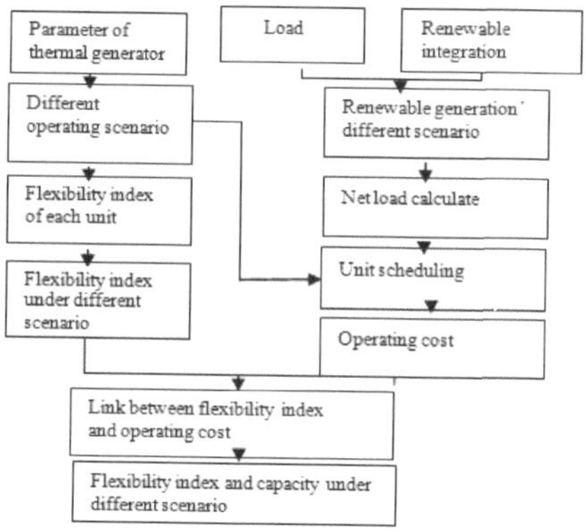

Fig. 1: Process for System Flexibility

Conclusion

The influence of generator flexibility on operational costs has gotten less attention. In order to deal with high fluctuation from renewable resources, a power system's flexibility is expected to increase. Furthermore, a typical generator's inflexibility increases device wear and tear, limits device lifespan, and increases maintenance costs. As a result, as renewable energy expands, more flexible generators will be necessary. One of the key topics of research in future power systems with a high renewable penetration is generating flexibility. Some example that give the today importance of generation flexibility, Lowering thc minimum generation level from 70% to 55% reduced VRE curtailment from 3.5 percent to 1.4 percent in India, China's flexible thermal power plant operation reduced VRE curtailment by 30%, and Germany's coal power plant refurbishment reduced minimum load by 170 MW and increased ramp rate by 10 MW per minute. Thermal generators must be flexible enough to allow for the cost-effective integration of significant variable renewable generating capacity. Due to a lack of reserve capacity, conventional thermal generators with poor flexibility have significant running costs. As a result, future flexible generator planning is required.

References

[1] Wu, Y., Tan, W., Huang, S., Chiang, Y., Chiu, C. and Su, C., "Impact of generation flexibility on the operating costs of the Taiwan power system under a high penetration of renewable power," in *IEEE Transactions on Industry Applications*, Vol. 56, No. 3, pp. 2348–2359, May–June 2020.

[2] Mohandes, B., Moursi, M.S.E., Hatziargyriou, N. and Khatib, S.E., "A review of power system flexibility with high penetration of renewables," in *IEEE Transactions on Power Systems*, Vol. 34, No. 4, pp. 3140–3155, July 2019.

[3] Impram, Semich; Nese, Secil Varbak and Oral, Bülent, "Challenges of renewable energy penetration on power system flexibility: A survey",

Energy Strategy Reviews, Vol. 31, 2020.

[4] Oree, Vishwamitra and Hassen, Sayed Z. Sayed, "A composite metric for assessing flexibility available in conventional generators of power systems," *Applied Energy*, Vol. 177, 2016.

[5] Guo, T., Gao, T., Zhou, X., Zhou., Li, Y. and Liu, J. "Optimal Scheduling of Power System Incorporating the Flexibility of Thermal Units". *Energies* 2018, 11, 2195.

[6] Anisie, Arina and Francisco Boshell. "Flexibility in conventional power plants: Innovation landscape brief." *Technical report, International Renewable Energy Agency (IRENA),* 2019.

[7] Taibi, Emanuele, *et al.* "Power system flexibility for the energy transition: Part 1, Overview for policy makers." 2018.

[8] Babatunde, O.M., Munda, J.L. and Hamam, Y., "Power system flexibility: A review", Energy Reports, Vol. 6, Supplement 2, 2020.

[9] Heggarty, Thomas; Bourmaud, Jean-Yves; Girard, Robin and Kariniotakis, Georges, "Quantifying power system flexibility provision", *Applied Energy,* Vol. 279, 2020.

[10] Papaefthymiou, Georgios; Haesen, Edwin and Sach, Thobias, "Power system flexibility tracker: indicators to track flexibility progress towards high-RES systems", *Renewable Energy*, Vol. 127, 2018.

[11] Roukerd, Saeed Poorvaezi; Abdollahi, Amir and Rashidinejad, Masoud, "Uncertainty-based unit commitment and construction in the presence of fast ramp units and energy storages as flexible resources considering enigmatic demand elasticity," *Journal of Energy Storage*, Vol. 29, 2020.

[12] Akrami, A., Doostizadeh, M. and Aminifar, F., "Power system flexibility: An overview of emergence to evolution," in *Journal of Modern Power Systems and Clean Energy*, Vol. 7, No. 5, pp. 987–1007, September 2019.

[13] Wind, load net load. "Flexibility in 21st Century Power Systems", *National renewable energy laboratory*, 2014.

[14] Min, C.-G. "Analyzing the impact of variability and uncertainty on power system flexibility. *Appl. Sci.* 2019.

[15] Emmanuel, Michael; Doubleday, Kate; Cakir, Burcin; Marković, Marija and Hodge, Bri-Mathias, "A review of power system planning and operational models for flexibility assessment in high solar energy penetration scenarios", *Solar Energy*, Vol. 210, 2020.

[16] Abujarad, S., Mustafa, M.W., Jamian, J.J., Abdilahi, A.M., De Kooning, J.D.M., Desmet, J. and Vandevelde, L. "An Adjusted Weight Metric to Quantify Flexibility Available in Conventional Generators for Low Carbon Power System's. *Energies* 2020.

[17] Heggarty, Thomas; Bourmaud, Jean-Yves; Girard, Robin, Georges Kariniotakis, "Multi-temporal assessment of power system flexibility requirement", *Applied Energy*, Vol. 238, 201.

Controlling and Monitoring of Feeders in Power System Distribution System using Supervisory Control and Data Acquisition (SCADA) System

Pappu Ram Bheel[1], M.K. Bhaskar[2] and Vinay Hans[3]

Department of Electrical Engineering (MBM Engineering College, Jodhpur)
Residency Road, Ratanada, Jodhpur–342011, India
E-mail: [1]princelot999@gmail.com; [2]mkb31@rediffmail.com;
[3]vinay.hans1310@gmail.com

Abstract: *SCADA is the acronym of Supervisory Control and Data Acquisition which means the communication scheme for telemetry measurement and control system used to monitor and remotely control the processes in utility industries such as electric power transmission and distribution, water distribution, and liquid and gas cross country pipelines [1].*

Scada is the network of physical devices embedded with electronics, software, sensors, actuators and network connectivity which have the ability to identify, collect and exchange the data. Future distribution systems will have to deal with an intensive penetration of distributed energy resources ensuring reliable and secure operation according to the smart grid paradigm. SCADA (Supervisory Control and Data Acquisition) is an essential infrastructure for this evolution [2].

Electric power system is a representative system using SCADA network allows objects to sensed or controlled remotely across existing network infrastructure, creating opportunities for more direct integration of physical world into computer-based systems and resulting in improved efficiency, accuracy and economy benefit with the added merits of minimum human intervention.

Keywords: SCADA; RTU; Real Time Data; Feeder; Controlling; Monitoring.

Introduction

Supervisory control and data acquisition (SCADA) systems are widely used to monitor and remotely control the processes in utility industries such as electric power transmission and distribution, water distribution, and liquid and gas cross-country pipelines. SCADA is being increasingly used in modern-day industries such as environmental monitoring, cell phone tower monitoring, well head monitoring, etc.

The benefits of SCADA systems are best appreciated when the process is very large—hundreds or even thousands of kilometers visits from one end to the other in terms of reducing the cost of routine to monitor facility operation. The value of these benefits grows even more if the facilities are very remote and require extreme effort to visit.

The Electricity Distribution Companies (EDC) normally identify the faulty section of the network and restore the power supply using their own resources which are mostly based on classical methods and techniques. Today the rapid growth of Information Technology (IT) tools has promoted many EDC to modernize their fault diagnosis as well as troubleshooting systems [1]. Among new technologies used for this purpose, SCADA systems are considered as the widely appropriate tool used for such processes. SCADA is the acronym for "Supervisory Control and Data Acquisition". SCADA systems are widely used for supervisory control and data acquisition of diverse kind of processes. Such process can be industrial, infrastructure or facility [3]. The SCADA system usually consists of the following subsystems [4]:

- A Man-Machine Interface (MMI) is the apparatus which presents process data to a human operator, and through this, the human operator, monitors and controls the process.
- A supervisory system, acquiring data on the process and sending commands to the process.
- Remote Terminal Units (RTU) connecting to sensors in the process, converting sensor signals to digital data and sending digital data to the supervisory system.
- Communication infrastructure connecting the supervisory system to the RTU.

In fact, most control actions are performed automatically by RTU or by programmable logic controllers (PLC). Host control functions are usually restricted to basic overriding or supervisory level intervention. For example, a PLC may control the flow of cooling water through part of an industrial process, but the SCADA system may allow operators to change the set points for the flow, and enable alarm conditions, such as loss of flow and high temperature, to be displayed and recorded [5–6]. The feedback control loop passes through the RTU or PLC, while the SCADA system monitors the overall performance of the loop. The main objective of this work is to show the need of the automation for EDC on their distribution networks and the importance of using computer-based system towards sustainable development of their services.

Basics of Scada System in Power System Distribution

A reliable power distribution system is an essential component for the economic growth and development of a country. Therefore, a modern electric power network system must be capable of performing 365 days a year and 24 hours a day with a high quality of uninterrupted power supply, even during the peak hours, to improve the performance of services to the customers. In view of the extensive size of the distribution networks, this can be achieved only by proper computer-based monitoring and control system as well as by efficient distribution and metering.

The "Monitoring and Control System" is the main part of a distribution automation network. This system was defined by IEEE as "A system that enables an electric company to remotely monitor, coordinate and operate distribution components in a real-time mode from remote location". The location from where control decisions are initiated is generally called Distribution Control Centre (DCC). Within this centre, different kinds of application software are used, which cooperate among themselves to achieve the control task [7–8].

SCADA refers to a system that collects data from various devices at factory, plant or in other remote locations like 33/11 KV Sub Station, 11 KV network and then sends these data to a control centre which then manage and controls the system.

SCADA focuses on gathering and circulating the right amount of system information to the right person or computer within the right amount of time. A simple approach to SCADA is shown in Figure 1.

Fig. 1: Simple Approach to SCADA

The typical SCADA system is a combination of four different subsystems:

1. The central computer (host system).
2. A field-located remote measurement and control equipment called.
   Remote terminal unit (RTU).
3. A wide-area communication system (telephonic, Internet, or intranet) to connect all these equipment. Typically, a fiber optic communication network is used to integrate the multiple pipelines and its related equipment as it is the most reliable medium for data transfer.
4. An operator interface that provides access to the system for the user.

A SCADA system enables an operator in a location central to a widely distributed process, such as, electric power transmission and distribution an oil or gas-filled, pipe-line system, irrigation system, or hydroelectric generating complex, to make set point changes on distant process controllers, to open or close valves or switches, to monitor alarms, and to-gather measurement information.

Typical SCADA/DMS Architecture

The architecture shown in Figure 2 illustrates a typical SCADA system in terms of the components or elements which are interconnected via a communication network.

In summary, there are five major areas or components to the SCADA system, as listed below and described in more detail:

- Field Devices and Signals
- Programmable Process Controller
- SCADA Operations User Work-station
- SCADA Server Computer
- Communication Network

The Field Devices are represented as signals into and out of the PPC; this includes controlling devices, such as pumps, valves, solenoids, etc. These field devices are the link between the SCADA technology and the process operations of any automated facility.

The PPC is the heart of any SCADA system, as the PPC performs the monitor and control functions over the field devices. Programming software for these components can become quite complex with multiple tasks, programs and routines.

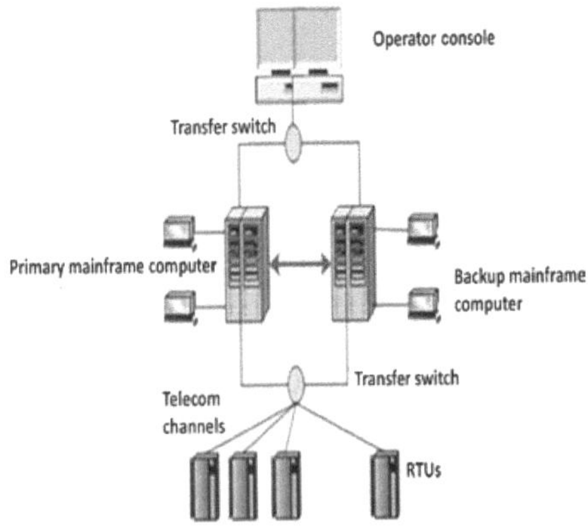

Fig. 2: Typical SCADA System Architecture

The SCADA Operations User Workstation provides the user interface in the form of process graphic displays, trends and associated reports. Most SCADA systems have multiple workstations through which the user can interact with the system to monitor operations and effect equipment control.

The SCADA Server Computer is used to maintain the process and historical databases for the user workstations, as well as provide the communication interface between the server computer and the PPCs.

The Communication Network is the hardware and software that intercom-nects all of the components of the system. Typical networks today include Ethernet with Transmission Control Protocol/Internet Protocol (TCP/IP) and proprietary communi-cation topology.

The SCADA/DMS system is designed to operate & monitor distribution substations and the network from a designed centralised location of each project area/town. The communication backbone for the envisaged SCADA/DMS system is MPLS-VPN and GPRS across the project area.

Elements of SCADA/ DMS System

SCADA systems are essentially process control systems (PCS) that gather, monitor, and analyse real-time environmental data from a simple power distribution system based on a predetermined set of data and conditions, such as consumed energy or power grid management. Some systems consist of one or more types of data acquisition devices such as remote terminal units (RTUs) and/or programmable logic controllers (PLC) that are connected to any number of field devices such as sensors, digital meters, protection relays, and station alarms (batteries, chargers and fire alarm).

SCADA systems are composed of the following equipment:

1. *Outstation Hardware:* Remote substation control devices, such as state of charge, current transformer (CT), voltage transformer (VT), fuel valves, and CB that can be controlled locally or remotely.
2. *Local Substation Processors:* Collect data from the field instruments and hardware equipment, including the RTU, PLC and intelligent electronic devices (IEDs), such as digital relays and digital meters. The local processor is responsible for dozens of analog and digital inputs/outputs from IEDs and switchgear equipment.
3. *Digital Instruments:* Usually installed in the field or in a facility and sense conditions such as current, voltage, irradiance, temperature, pressure, wind speed, and flow rate.
4. *Communications Devices:* Could be either short-range or long-range Communications Short-range communications are installed between local RTUs, instruments, and operating equipment. These are relatively short distance cables or wireless connections that carry digital and analog signals using electrical properties, such as voltage and current, or other settled industrial communications protocols. Long-range communications are installed between local processors RTU/PLC and host servers. This communication typically uses methods such as leased telephone lines, microwave, satellite, frame relay network, and cellular packet data.
5. *Host Computers/Servers:* Host Computers, such as data acquisition computer server, engineering/operation workstations. As the central point of monitoring and control, they will be in the control room or master station. The operation workstation is

where an engineer or operator can supervise the process, as well as receive system alarms, review data, and exercise remote control.

Following the illustration are more detailed explanations of each of major components.

Remote Terminal Unit (RTU)

A Remote Terminal Unit (RTU), with panel in Figure 3, is a Microprocessor Based Electronic device that interfaces objects in the physical world the Field (Sub-Station) to SCADA Centre, by transmitting data to the system, and by using messages received from the Control Centre to control connected Field objects. Another term that may be used for RTU is remote telemetry unit the common usage term varies with the application area generally.

Fig. 3: RTU (Remote Terminal Unit) Panel

LDMS Introduce

A Local Display Monitoring System (LDMS) or human machine interface (HMI) is the apparatus which presents process data to a human operator, and through which the human operator controls the process. HMI is usually linked to the SCADA system's databases and software programs, to provide trending, diagnostic data' and management information such as scheduled maintenance procedures, logistic information, detailed schematics for a particular sensor or machine, and expert-system troubleshooting guides.

Basic of Feeder Remote Terminal Unit: A Feeder Remote Terminal Unit (FRTU) is a Microprocessor Based Electronic device that interfaces objects in the physical world the Field (RMU) to SCADA/DMS Centre by transmiting data to the system, and by using messages received from the Control Centre to control connected Field objects.

FRTU (Feeder Remote Terminal unit) Figure 4 is installed on distribution pole and controls (connect/disconnect) electricity flow on distribution line in remote place through wire or wireless communication. It helps reliable and efficient distributed electric line operation to be optimized with supervising supply of electricity. FRTU installed in the field is divided into the processing for interfacing with the switching device to the switching power distribution line and the interfacing for interfacing with the switching device for the underground distribution line.

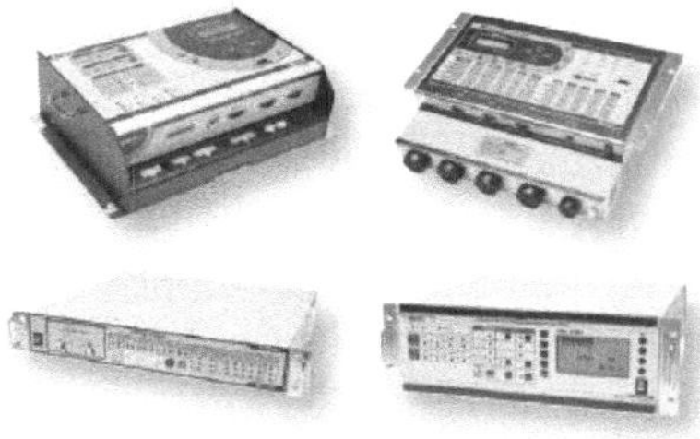

Fig. 4: Feeder Remote Terminal Unit

The main function of the FRTU is to measure the current and voltage flowing in the distribution line in real time through the CT and CPD built in the switchgear, detect the condition at the time of the accident with the power distribution line, transmit to the central control system, and performs a remote-control function for the opening/closing device. In addition, it stores all events occurring in the distribution line to provide an analysis of the cause of the accident, and transmits the status of the distribution line to the central device using the DNP protocol to perform remote monitoring of the condition and control of the load switching device.

Controlling of Feeders using SCADA

Many other types of equipment will also be used to support such automation of a power network. They include Automatic Meter Reading (AMR), Data Concentrator Unit (DCA), Remote Terminal Unit (RTU), Supervisory Control and Data Acquisition (SCADA), Communication equipment, etc. In this discussion we would mainly focus on the customer service quality. Customer service requirements point to one key element: Information, i.e., the right amount of information to the right person or computer within the right amount of time. The flow of information requires data communication over extended networks of systems and users. In fact, utilities are becoming among the largest users of data and are the largest users of real-time information [9].

Hence, the implementation of power network automation system will provide better services to EDC customers and improve the power quality and reliability of the electric supply services, which would satisfy the following goals:

- Respond to customer service interruptions more quickly.
- More efficiency of the power system by maintaining acceptable power factors and reduced losses.
- More control and limit of peak power demand.
- Ability of EDC engineering staff to monitor and control the power system during normal and abnormal conditions by providing more reliable and appropriate real time data.
- Ability of EDC engineering staff to perform the power system analysis and planning by providing increased. access to past and current operations data and associated software tools.

- Ability of EDC engineering staff to manage the power system assets and system operations by providing increased access to better performance data and historical statistics.

The terminal units at the customer location will be used for many tasks. They will display the consumption, service and billing information including hourly load curves. The customer can also remotely close as the circuit breaker (as in Table 1) of an outdoor meter equipped with the motor option. Also, the terminal units will communicate with the control centre to receive messages, alert the client in time to reduce the load when it exceeds a given limit and prevent tripping of the meter breaker and finally support prepayment system.

Table 1: Reasons of Outage of Feeder by SCADA

S. No.	*Sub-station Name*	*Feeder Name*	*Tripping Date Time*	*Closing Date Time*	*Outage Time*	*Reason of Outage*
1.	P	A	23-04-2020 09:05	23-04-2020 09:17	00:11:29	Earth Fault
2.	P	B	23-04-2020 11:12	23-04-2020 11:22	00:10:20	Manually
3.	P	C	24-04-2020 14:36	24-04-2020 14:42	00:05:51	Manually
4.	P	C	25-04-2020 08:00	25-04-2020 08:13	00:13:05	Earth Fault
5.	Q	A	24-04-2020 06:54	24-04-2020 07:23	00:28:48	Over Current
6.	Q	A	25-04-2020 08:49	25-04.2020 09:35	00:46:25	Earth Fault
7.	Q	A	25-04-2020 11:35	25-04-2020 11:41	00:05:41	Earth Fault
8.	R	A	25-04-2020 12:41	25-04-2020 12:53	00:11:48	Manually
9.	R	A	26-04-2020 17:17	26-04-2020 17:29	00:12:53	Over Current
10.	R	A	27-04-2020 07:26	27-04-2020 09:55	02:29:15	Manually
11.	R	A	27-04-2020 16:55	27-04-2020 17:15	00:20:08	Earth Fault

Following are the main benefits of SCADA control system:

- Save time and money.
- Reduces man-power needs.
- Increases production efficiency of a company.
- Cost effective for power systems.
- Reliable.

Conclusion

The paper presents a SCADA system for a distribution system. All Sub-station are delivering power to consumers demand having Supervisory control and data acquisition (SCADA) systems are widely used to monitor and remotely control the distribution. This paper provides a way to include supervisory control and Data Acquisition of whole power Distribution network as well as the power consume by the consumers. The main advantage of this paper is that it provides un-interrupted power supply to consumer. The additional advantage is that there is a monitoring with controlling. So authorized user may change the control settings according to requirement. Beside this by implementing SCADA control system we can save time and money, reduced man-power needs, increase production efficiency of a company.

References

[1] Lakhoua, Mohamed Najeh and Jbira, Mohamed Kamel, "Project Management Phases of a SCADA System for Automation of Electrical Distribution Networks", *IJCSI International Journal of Computer Science Issues,* Vol. 9, March 2012.

[2] Vale, Zita; Morais, Hugo; Faria, Pedro and Ramos, Carlos, Distribution system operation supported by contextual energy resource management based on intelligent SCADA.

[3] Bailey, D. and Wright, E., "Practical SCADA for Industry", Elsevier, 2003.

[4] Clarke, G., Reynders, D. and Wright, E., "Practical Modern SCADA Protocols", Elsevier, 2003.

[5] Wiles, J., "Techno Security's Guide to Securing SCADA: A Comprehensive Handbook Protecting The Critical Infrastructure", Elsevier, 2008.

[6] Chan, E.K. and Ebenhon, H., "The implementation and Evolution of a SCADA System for a Large Distribution Network," *IEEE Transactions on Power systems*, Vol. 7, No. 1, 1992.

[7] Bassett, D., Clinard, K., Grainger, J., Purucker, S. and Ward, D., Tutorial Course: Distribution Automation, *IEEE Publication*, 1988.

[8] Becker, D., Falk, H., Gillerman, J., Mauser, S., Podmor, R. and Schneberger, L., Standards-Based Approach Integrates Utility Applications, *IEEE Computer Applications in Power*, Vol. 13, 2000, pp. 13–20.

[9] Grasberg, L. and Osterlund, L.A., "SCADA EMS DMS-a part of the corporate IT system", IEEE Power Engineering Society International Conf., 2001.

Compact Fluorescent Lamp Harmonic Analysis and Mitigation using Passive Filter

Mulualem Tesafye[1], Baseem Khan[2] and Om Prakash Mahela[3]

[1,2]Department of Electrical and Computer Engineering,
Hawassa University, Hawassa, Ethiopia, 05

[3]Power System Planning Division, Rajasthan Rajya Vidyut Prasaran
Nigam Ltd., Jaipur–302005, India

E-mail: [3]opmahela@gmail.com

ABSTRACT: *These instructions give you the basic guidelines for Modern electrical devices such as Compact Fluorescent Lamp (CFL), many of which are nonlinear loads inject harmonics into the power Distribution Network. The current and voltage Total Harmonics Distortion (THDi and THDv) being emitted by CFL are within the range of 90%–200% of the fundamental current and 20%–40% of the fundamental voltage respectively. These harmonics cause power quality problems such as overheating of Distribution Transformer (DT) and low tension conductors, mal-operation of energy meter, deterioration of protective device. Mitigation of these harmonics is harder once injected into the network due to their distributed nature. Consequently, this research work modeled and simulated compact fluorescent lamp for the analysis of harmonic injection into power distribution network and the effect of harmonic distortion produced by 3000 lamps that is found around IOT student dormitory on PDN is also analyzed. A suitable method was also developed for mitigating the harmonics emissions before they are injected into the distribution network with the view to mitigating its effects. Linear Technology Spice (LTspice) software package was used to generate voltage waveform and current waveform. The results obtained for THDi and THDv were in relation to the CFL fundamental current and voltage are 117.782% and 31.324% LTspice software. These results were found to be violating the Institute of Electrical and Electronics Engineers (IEEE) standard limits for harmonics. Passive shunt filter networks were also developed to mitigate the harmonics. CFL Simulink models with shunt passive harmonics filters networks were developed.*

Keywords: Compact Fluorescent Lamp (CFL), Switch-Mode Power Supply (SMPS), Industrial Power Corruptor (IPC), Total Harmonic Distortion (THD).

Introduction

In recent times, there has been a widespread increase in thc number of compact fluorescent lamps (CFLs) connected to the electrical power distribution network. From the standpoint of the energy company, understanding the effects of such technologies on distribution systems is critical. A compact fluorescent lamp is a subset of a compact fluorescent lamp. CFLs are meant to replace incandescent lamps (IL) and may be used in most current light fixtures, even those that were built for incandescent lamps. This is due to the advantages of cost-efficient, energy-efficient, and longer-lasting compact fluorescent lamps (CFLs) over incandescent lamps [1].

Due to the above-listed advantages of CFLs, the lamps are being promoted by governments and electricity supply utilities worldwide as part of energy conservation programs. CFLs are rapidly being employed in both household and business settings, substituting a vast number of incandescent lights and other lighting systems [2].

One of the most significant disadvantages of CFLs is that they increase the degree of disruption that may harm both customers and utility devices on the power system network. Because it produces and inserts harmonics into the power distribution system, this is the case. A sinusoidal element of a sinusoidal waveform with a frequency that is an integral multiple of the electricity system's fundamental frequency is referred to as a power network harmonic. In contrast to IL, which operates through heating, CFL operates through chemical reaction. Every CFL uses an electronic ballast (EB) circuit to generate a high frequency within the range of 10 kHz to 40 kHz, which in turn enables the tube to emit visible light from its inner coating. In other words, the EB of the CFL uses a rectifier to input power to an inverter, which in turn supplies the necessary AC power to the lamp. The phenomenon of the ballast circuit is the major source of harmonics in CFLs. It is responsible for a rise in harmonic current injections into the electrical power distribution network.

Some previous studies on compact fluorescent lamp CFL modeling for harmonic analysis used laboratory test beds to measure the harmonic emissions of individual or grouped CFLs in order to provide a platform for harmonic analysis. This method is associated with the challenge of cut-off of harmonics being injected into the lamp by the distribution network through the AC current being drawn by it. This affects the accuracy of the overall results. A more appropriate procedure for modeling the lamp is required in order to clearly quantify the harmonics produced by CFLs precisely. Other researchers carried out harmonic analysis in the frequency domain. In the method of harmonic analysis in the frequency domain, the principle of superposition is applied to enable each harmonic to be considered separately. The differential equations thus obtained are usually converted to complex algebraic equations. These equations are then represented by complex phasors, either in rectangular or polar form. Even though this method was said to have the advantage of computational efficiency, it is complex. In light of the foregoing, this study used software LTSpice to design and simulate compact fluorescent lamps for the assessment of harmonics injection into power distribution networks and established a suitable method of harmonics emission reduction with the goal of reducing their injection into the power distribution system.

Higher frequencies that are undesired are overlaid on the basic waveform, resulting in a deformed wave form. A perfect AC generator produces a pure sinusoidal voltage as a conceptual quantity. The operation of non-linear loads produces harmonic frequencies. They are a common source of power quality issues, including increased device and conductor over-heating, variable speed drive malfunc-tioning and motor and generator torque vibrational modes.

A switch-mode power supply (SMPS) is a power converter that employs switching devices such as MOSFETs, which switch on and off at a high frequency, as well as energy storage systems such as capacitors and inductors, to provide power during the switching devices' non-conduction states.

A system of electrical elements used to provide, transport, and consume electric power is referred to as a power grid. Energy is provided by some types of production; the transmission and distribution are done by a transmission and distribution system; and the con-sumption is provided by the domestic sector. But the nonlinear loads, such as CFLs, that are found inside residential houses introduce harmonic distortion into the power system. The combined effects of the wide use of CFLs can be just as large as the harmonic source. Moreover, mitigation of harmonic distortions caused by nonlinear loads such as the CFL is very difficult once they are injected into the power distri-bution network [4]. Excess harmonics can lead to disturbances in electricity supply, reduced power quality, malfunctioning of components of a power system, and line overheating. If severe harmonics occur in inter-connected loads, the harmonics cause the electrical device (or load) to malfunction, fail prematurely, or not operate at all. Moreover, this harmonic is created due to load expansion in the power grid, and it will cause fatal damage to the network that is connected to the grid and the transmission components. Therefore, this research modeled and simulated compact fluorescent lamps for the analysis and mitigation of harmonic emissions before they are injected into the power distribution network.

Related Work

Some fundamental concepts related to this research work are reviewed. Some of the concepts. The topics discussed here are power system harmonics, power factor, support model equations relevant to the research and other aspects of significance to the research.

Mohamed *et al.* conducted a study on harmonics from compact fluorescent lamps. The research investigated the amount of harmonic generation from CFLs. This was achieved in a laboratory test bed using an Industrial Power Corruptor (IPC). The result of the study proved that CFLs produce harmonics as a result of their operation and increase the value of current THD for this type of electrical load. However, the results obtained are largely dependent on the value of the current measurement resistor (CMR). Consequently, the integrity of the result may be affected when a large current greater than the limiting current of the CMR is drawn by the lamp [6]. Teodosescu *et al.* conducted a study on the improvement of a CFL's power factor by using a valley fill filter. The research analyzed a method of improving a compact fluorescent lamp's power factor through the use of two passive valley filters. The circuit diagram of the 18W Philips ecotone was put into the Orcad program and simulated by the computer. According to the researcher, the current total harmonic distortion was found to be 77% and was reduced to 66% and 29%. However, none of the results complied with the IEEE standard limit of 20% [6]. A study on the model of electronic ballast compact fluorescent bulbs was provided by Molina *et al.* Depending on the CFL equivalent circuit, the study constructed a CFL model. The CFL models, as well as a simple estimate technique for obtaining model parameters from experimental observations, were utilized to calculate their harmonic current injections into the distribution network. The study suggested a simple approach for estimating the electrical properties of the developed CFL equivalent circuit based on CFL supply voltage and ac current measurements. Both the proposed model and the estimation algorithm were validated with laboratory tests. The significance of the model developed was limited to the prediction of harmonic current emission by CFLs with large ac resistances and small dc capacitors. However, a model is required that can predict the emissions of any CFL into the distribution network. Pileggi *et al.* carried out research on the effect of modern compact fluorescent lights on voltage distortion. The study was one of the pioneers in the research area. It considered and measured the harmonic contribution of CFLs with low input power of 9 W to 33 W, which were used in a residential area of 3,000 households. The work established that the CFL being used in the residential area would create a total load of 300 kW to 500 kW. The researchers employed a computer simulation of three real-life 13.8 kV feeders feeding nonlinear demands such as CFLs with electronic ballasts to conduct their research. Laboratory experiment results and CFL installation/ monitoring in the house backed up the computer calculations. The information gathered was utilized to create load models. According to the research, the input current to the CFL has an abnormally high total harmonic distortion, ranging from 116 percent to 142 percent, according to the research. According to the findings, the total load of CFLs should not exceed 100 kW for the safe operation of a 15 kV feeder with a maximum 10 MVA load capacity. However, at some locations along the feeder, this might result in intolerable voltage distortion. A 5% voltage distortion for a few minutes every day may be tolerated and have no effect on the distribution network, but the same distortion over several hours may produce system damage. Design and modeling of a shunt passive filter for non-linear load harmonic mitigation, K.M. John *et al.* The paper presents a study of the shunt passive filter design procedure to mitigate harmonics of non-linear loads. The non-linear load considered in this paper is DC fed [7]. Filters are designed by M. Kevin *et al.* to eliminate harmonic distortion in industrial power systems. Numerous three-phase filter banks are employed and linked in parallel to produce acceptable distortion, improve power quality, and eliminate harmonics. High order harmonic situations have been suppressed in this thesis using Butterworth, Chebyshev, and Cauer filter versions [1]. D. Leva *et al.* did research on end-user device power quality and harmonic analysis. The voltage and current waveforms were measured using a 300 MHz, four-channel digital oscilloscope in the study. The research investigated the effect of CFLs on power systems in relation to harmonic distortions and low power factors inherently associated with CFLs. The study established that the lamps generate substantial harmonic distortions and lead power factors. However, it proposed that the combination of IL and CFLs in one circuit would improve the power quality of the distribution network. This method of harmonic analysis does not take into consideration the idea of energy saving, as using ILs to reduce the harmonics will result in energy wastage. Rawa *et al.* did a study on FL and CFL harmonic investigations using experimental measurements and computer simulations. The study created a model to quantify the harmonics generated by FLs and CFLs. To get close to the experimental values, MATLAB Simulink was used to construct and simulate the entire models of FLs and CFLs. A Constant Load Power (CLP) model was also created, with the results compared to the complete FL and CFL models. According to the researchers, it is faster and easier to simulate the CFL 32 model using the CLP model than the full model. However, using the developed CLP model, any variation in voltage may affect harmonics generated by the CFLs. As a result, the CFLs must be powered by chromaTM, a programmable AC source that isolates the test rig while filtering out harmonics and fluctuations from the main power source. The model does not have universal application. In much of the literature reviewed, harmonics were the utmost concern in terms of the power quality of the distribution network while using CFLs. However, previous studies in harmonic analysis were marred by the following flaws.

Do not provide optimal harmonics injecttion mitigation.

- Do not have universal application.
- Complexity.
- Required large computational efforts.

In light of the aforementioned difficulties, this study modeled and simulated a small fluorescent bulb for the purpose of analyzing harmonics injection into the supply system. LTspice software was used to develop the CFL equivalent. The LTspice was also used to perform an FFT harmonic analysis on the CFL equivalent circuit.

The general objective of the project is to analyze harmonic distortion caused by CFLs and design filters to reduce harmonics produced by these loads in order to reduce the effect they cause on the power system. The purpose of the research is to reduce CFL current and voltage harmonics injections into power distribution networks to conform to IEEE standard limits. The contribution of this work is,

- To analyze voltage and current harmonic distortion.
- Develop CFL model and determine its current and voltage total harmonics distortions being injected into power distribution network.
- Develop harmonics filter network for the CFL harmonics injections mitigation.
- To compare results obtained from the simulation without filter and with filter.
- Validate results obtained through comparison with measured values from experimental fixed bed.

Methodology

In order to achieve the main aim of the project, there are various procedural tasks we follow. The first method toward the processing of the work starts with reviewing different literature where all the theoretical information regarding harmonic distortion is available. This is followed by studying the characteristics and modeling the non-linear load, filter design, and developing CFL on LT Spice. Finally, total harmonic distortion before adding the filter to the circuit and after that is computed.

Furthermore, harmonic distortion after the filter is inserted is analyzed and compared to the previous result. Figure 1 presents the summary of the methodology.

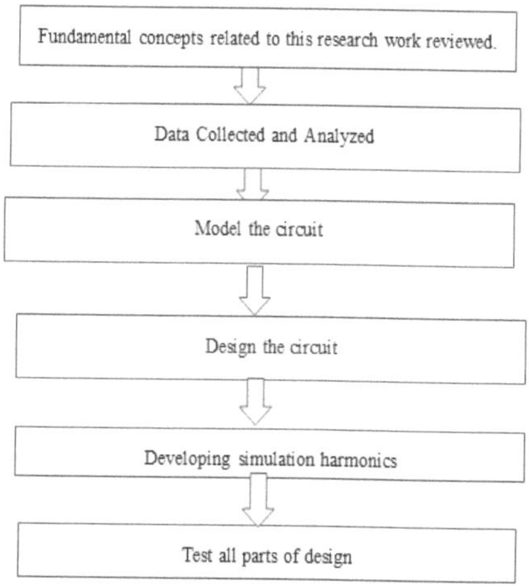

Fig. 1: Summary of Methodology

Block Diagram Description

The system considered for harmonic analysis here consists of a single phase

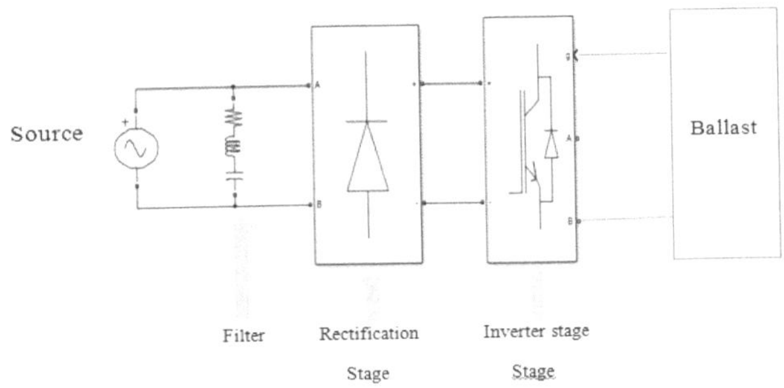

Fig. 2: Block Diagram of CFL

220 V, 50 Hz AC source which feeds AC nonlinear load through a rectifier and inverter setup. The rectifier consists of a four diode which performs un-controlled rectification and inverter which convert the Ac voltage in to DC. This setup is of nonlinear load 1. This rectifier and inverter consist of power electronic switches which inject harmonics on the AC source side because of their switching characteristics. This harmonic created due to the switching devices is mitigated by using harmonic filters. Figure 2 presented the block diagram of CFL.

Harmonic Analysis using Fourier Transform

In order to comprehend the distortion phenomenon and give relevant methods for its mitigation, harmonics analysis of the distorted waveform is required. Harmonic analysis allows the distorted waveform to be expressed as a sum of the DC part, the deformed waveform's basic sine wave, and a succession of pure sine waves. Sine waves have varying magnitudes, and their frequencies are integer multiples of the basic distorted waveform, which is represented by the Fourier series. The following equations can be used to study a distorted waveform using Fourier's series.

$$f(x) = a_0 + \sum_{n=1}^{\infty}(a_n \cos(\mathrm{nwt}) + b_n \sin(nwt)) \quad \text{... (1)}$$

$$a_0 = \frac{1}{T}\int_0^T f(t)d(t) \quad \text{... (2)}$$

$$a_{n} = \frac{2}{T}\int_0^T f(t)cos(nwt)d(t) \quad \text{... (3)}$$

$$b_{n} = \frac{2}{T}\int_0^T f(t)sin(nwt)d(t) \quad \text{... (4)}$$

where, n = 1, 2, 3 …, n $f(t)$ is the function in time domain, ω is the fundamental frequency, a_n and b_n are Fourier cosine and sine coefficients respectively. The Fourier series has only odd harmonic components when the positive and negative half cycles of a waveform have the same forms with respect to the vertical axis and the average value of its period is equal to zero.

Total Harmonic Distortion (THD)

The cumulative degree of distortion within an electrical current in comparison to the ideal is known as total harmonic distortion (THD). The most practical approach for determining total harmonic distortion for current signals and total harmonic distortion for voltage signals is total harmonic distortion (THD). The ratio of the RMS values of all harmonic signals to the RMS value of the fundamental signal frequency is known as THD.

Current THD

As per the above statement total distortion for current is indicated by THDi,

$$\mathrm{THDi} = \frac{\sqrt{\sum_{2}^{\infty} \mathrm{In}^2}}{\mathrm{I1}}$$

Here, in is the RMS current for the nth harmonic signal and I1 is the RMS value of the fundamental signal.

Voltage THD

Same as THDi, total harmonic-distortion of voltage is denoted by THDv.

$$\mathrm{THDi} = \frac{\sqrt{\sum_{2}^{\infty} \mathrm{Vn}^2}}{\mathrm{V1}}$$

The voltage of the nth harmonic is Vn, whereas the voltage of the fundamental signal is V1. With the Fast Fourier transform, total harmonic distortion (THD) evaluates the system's nonlinear behavior (FFT).

Mathematical Design and Simulation

Mathematical Design

For the investigation, a circuit schematic of an actual CFL 15 W half spiral, white in color, pin type CFL was employed. The CFL was numbered MD HS-15 W B22 (220–240 V 50/60 Hz 0.145 A) which was a product of YESIMODI, a company in China. The CFL was made in China. Lifespan of the lamp was put at 8,000 hrs. for 4 hrs. operation per day by the manufacturer. More over all values of the components available in the circuit diagram of CFL are the values that the company used to develop the lamp.

Shunt Passive Filter Design

A passive filter is made up of capacitors and inductors that are set to resonate at a single frequency or a range of frequencies. The absolute value of capacitance, inductance and the series resistance of the nth order filter were calculated using the following equations,

$$C_{n} = \frac{Qc}{m\omega Vs^2} \quad \text{... (5)}$$

$$L_{n} = \frac{1}{n^2\omega^2 Cn} \quad \text{... (6)}$$

$$R_{n} = \frac{n\omega Ln}{Qn} \quad \text{... (7)}$$

where,

$$Qc = \sqrt{S^2 - P^2} \quad \text{... (8)}$$

$$\mathrm{S} = \mathrm{VI} \quad \text{... (9)}$$

$$\mathrm{P} = \mathrm{SCos}\theta \quad \text{... (10)}$$

Here, m is the number of branches in the passive filter for all the phases, V_S is per-phase fundamental voltage across the filter, ω is the fundamental frequency of the supply, Qn is the quality factor of the resistor and it is measure of the performance of a resistor of the nth order filter, which is normally considered as $10 < Q < 100$.

Using the following equations, the apparent (S), real (P) and reactive (Qc) powers of the CFL were 31.28 VA, 17.20 W, and 26.12 VAr respectively.

Also, the Capacitance of the Capacitor (Cn) used to develop the filters is 0.8 μF with ω is 314.2.

Results and Discussion

This section presents and discusses FFT harmonic analyses results obtained before and after insertion of the developed harmonics filter networks. Simplified FFT algorithm, LTspice software was used for the harmonics analyses. Technology Spice software IV is utilized to draw the circuit. The CFL circuit diagram in Figure 1 was developed on the LTspice in order to obtain current and voltage waveforms of the signal to be used for the FFT analysis. The circuit diagram of single CFL is given in Figure 3.

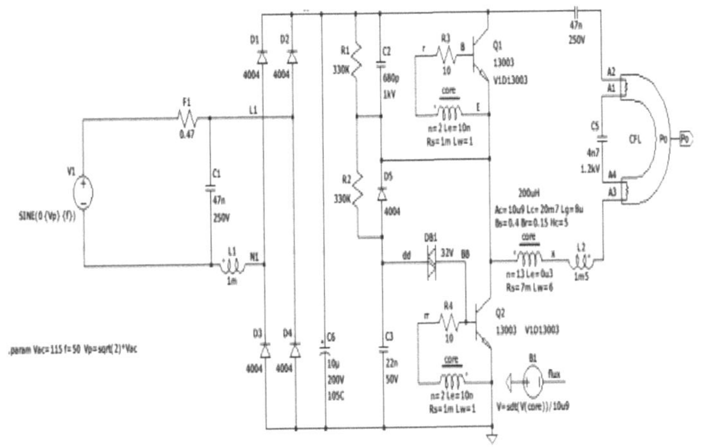

Fig. 3: Circuit Diagram of Single CFL

CFL Voltage Waveform

The voltage waveform of the signal of the replicated CFL circuit on LTspice is given in Figure 4.

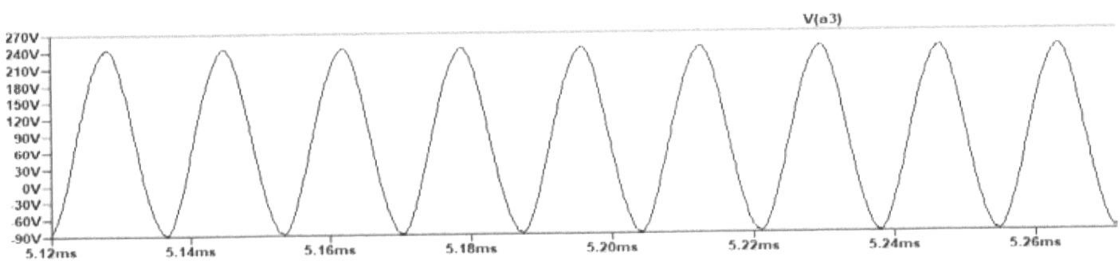

Fig. 4: CFL Output Voltage Waveform

CFL Current Waveform

The CFL current waveform without developed filter on LTspice software is as shown in Figure 5.

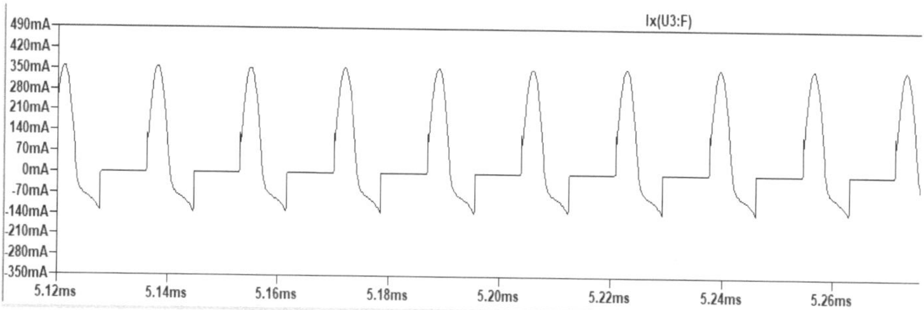

Fig. 5: CFL Current Output Waveform

Here, also the current waveform shown in Figure 5 is none smooth cosine waveform. In addition, the current waveform has a DC element. The non-smoothness of the waveform reveals the existence of harmonics. The distorted source current and voltage due to switching device that the circuit use to get high frequency output is as shown in Figures 6 and 7.

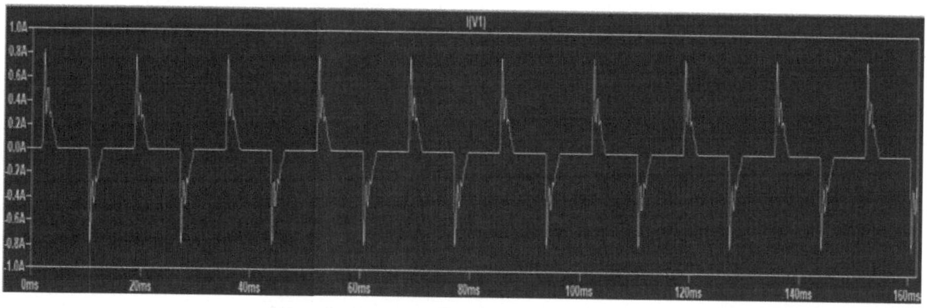

Fig. 6: Source Current Wave Form

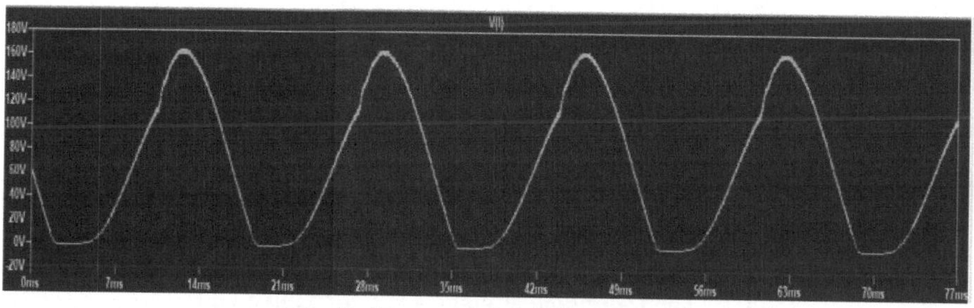

Fig. 7: Source Voltage Wave Form

As we observe from above figures the current and voltage waveform shown is none smooth cosine wave-form. A DC component is also included in the current waveform. Due to the waveform's non-smoothness, the existence of harmonics is visible.

LTspice FFT Harmonics Analysis Results

Results obtained when FFT harmonics analysis was carried out on the CFL with LTspice were 117.78% and 31.32% for THDi and THDv. The THDi and THDv obtained were far greater than the IEEE 519-2019 harmonics standard limits of 20% and 8% respectively. The results of the spectrum analysis for current and voltage are as shown in Figures 8 and 9 respectively. The Current THD for single CFL can be drawn using Bar graph as shown in Figure 8.

The voltage THD for single CFL can be drawn using Bar graph as Figure 9.

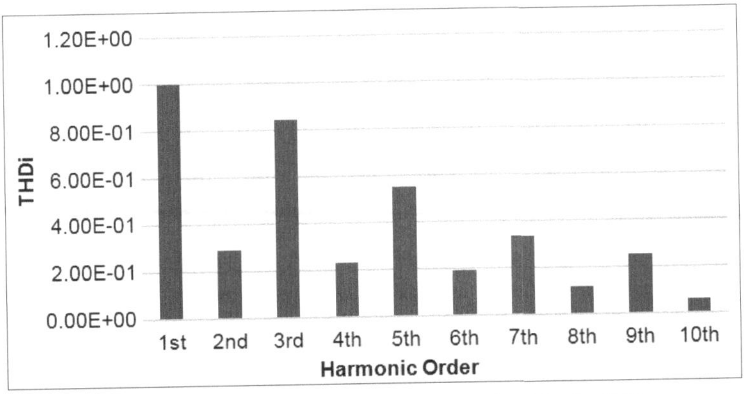

Fig. 8: The Individual Current THD for Single CFL

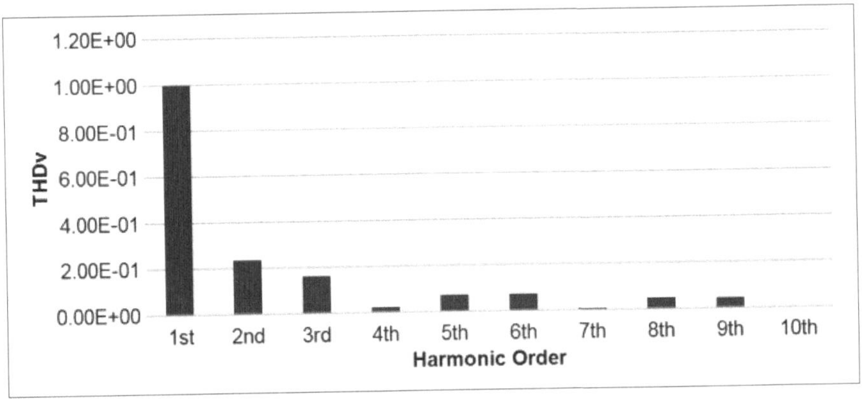

Fig. 9: The Individual Voltage THD for Single CFL

Harmonics Measurements of Multiple CFL

CFLs are rapidly being employed for a variety of purposes in the industrial, commercial, and residential sectors, because the loads take distorted current from the source, the supply voltage at the point of common coupling is significantly distorted (PCC). The CFLs located near Hawassa University Institute of Technology's student housing were used as a case study. Individual and combined CFL loads' harmonics are investigated. In the dormitories, there are around 3000 CFL. These CFLs create non-sinusoidal wave shapes as well as odd and even order harmonic frequencies that are multiples of the fundamental frequency. When this harmonic current passes via cables, transformers, and circuit breakers, it causes distorted voltage at different frequencies to reach the PDN.

Two CFL

The circuit while connecting two CFL together is given like Figure 10.

Results obtained when FFT harmonics analysis carried out on LTspice were 81.397% and 0.2112% for THDi and THDv respectively. The current THD for two CFL can be drawn using Bar graph as Figures 11 and 12.

The voltage THD for two CFL can be drawn using Bar graph as Figure 12.

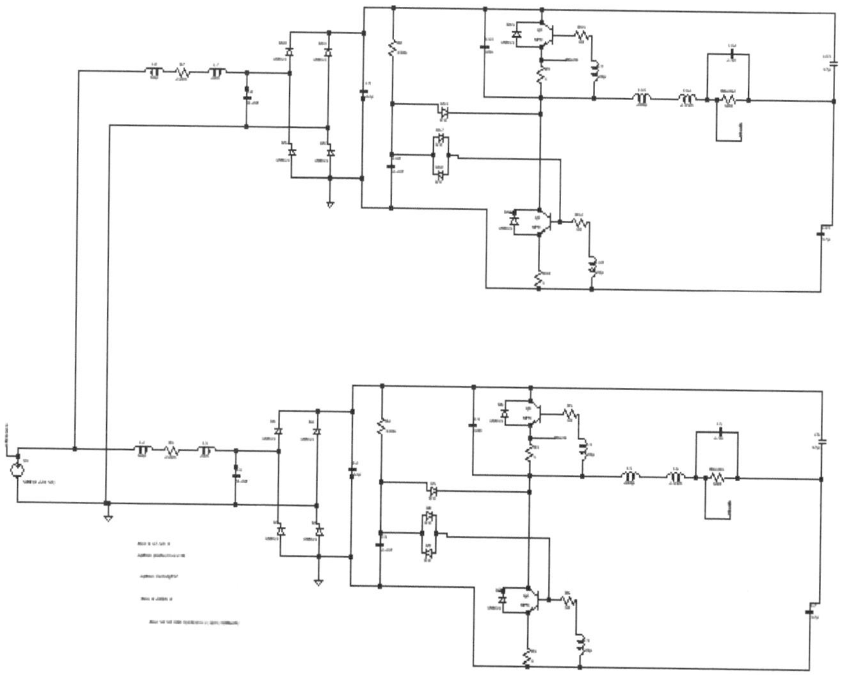

Fig. 10: The Circuit Diagram for Two CFL

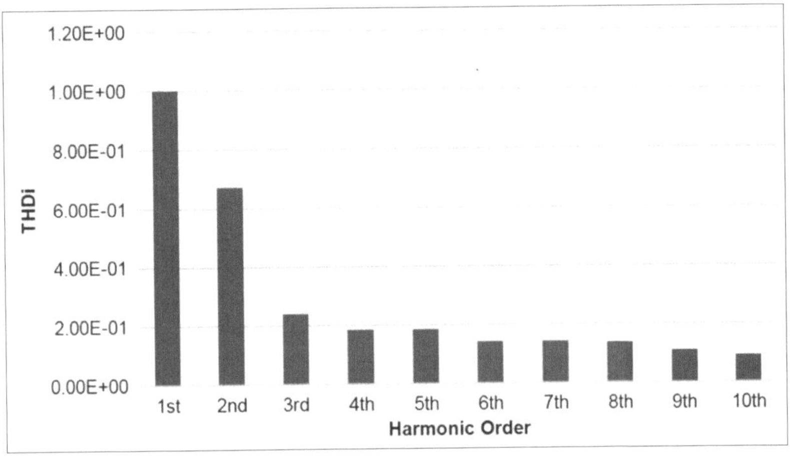

Fig. 11: The Individual Current THD for Two CFL

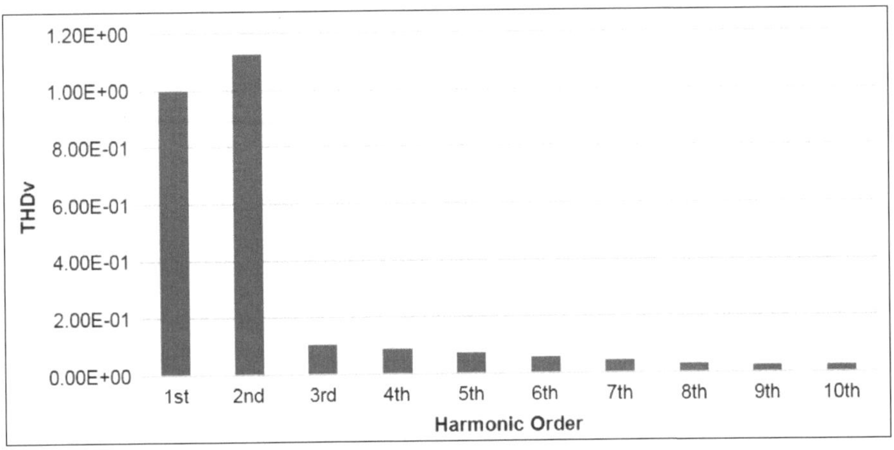

Fig. 12: The Individual Voltage THD for Two CFL

Four CFL

The circuit while connecting four CFL together is given like Figure 13.

Results obtained when FFT harmonics analysis carried out on LTspice for four CFLs connected in parallel were 82.748% and 0.2402% for THDi and THDv respectively. The voltage THD for two CFL can be drawn using bar graph as Figures 14 and 15.

The current THD for four CFL can be drawn using Bar graph as Figure 15.

As we tried to show above by connecting many CFL together the harmonic Distortion we obtained from FFT of LTspice as we increase number of CFL is shown in Table 1.

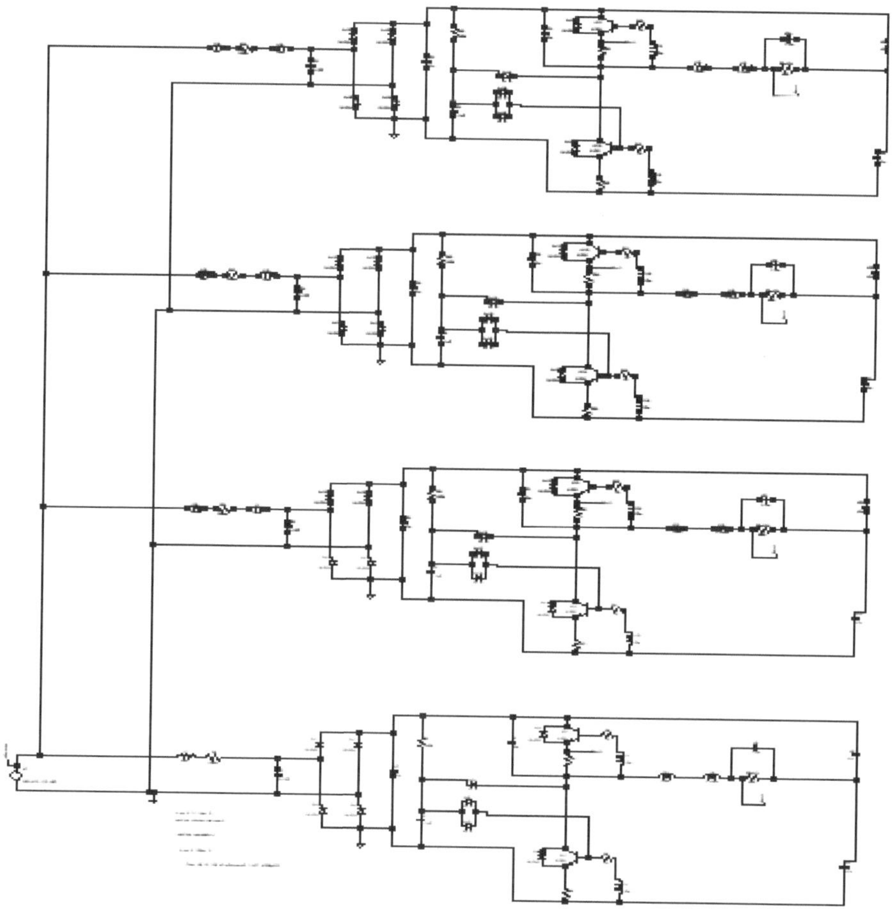

Fig. 13: The Circuit Diagram for Four CFL

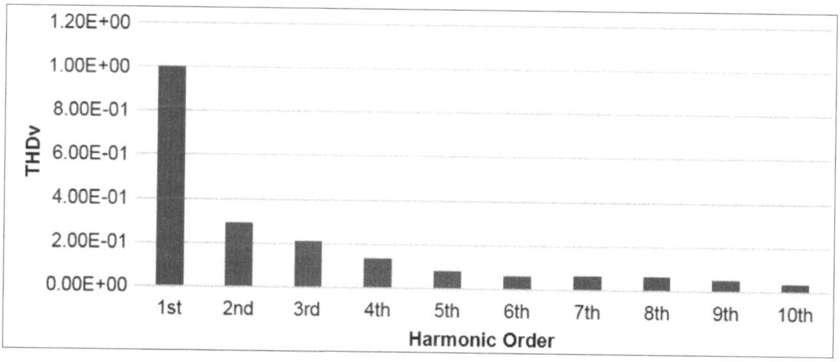

Fig. 14: The Individual Voltage THD for Four CFL

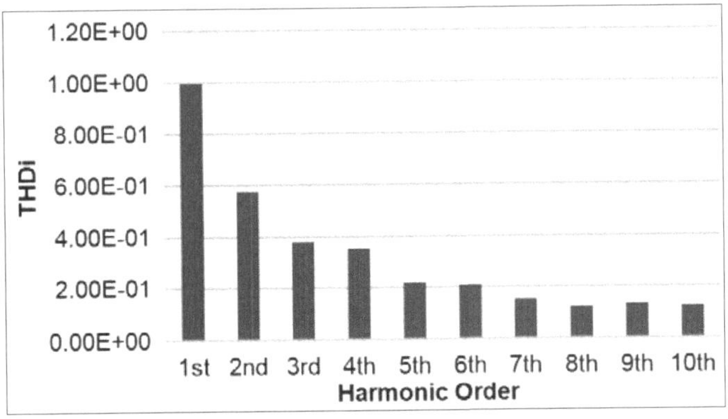

Fig. 15: The Individual Voltage THD for Four CFL

Table 1: THDi *vs.* Number of CFL

Number of CFL	*THDv*	*THDi*
2	21.2112	81.395
4	24.328	82.7454
6	27.664	83.764
8	31.0942	84.568
10	34.559	85.309
14	41.497	87.3021
20	49.347	95.942

But in our case, we have taken the total number of CFLs that are found on IOT dormitory is 3000 and to obtain the THDi and THDv caused by these CFLs we estimated based on above data using linear regression method like shown in Figure 16 by plotting the THD *vs.* number of CFL on excel. After plotting the graph we drive equation using linear regression. Figure 16 presented the THDi as number of CFL increase.

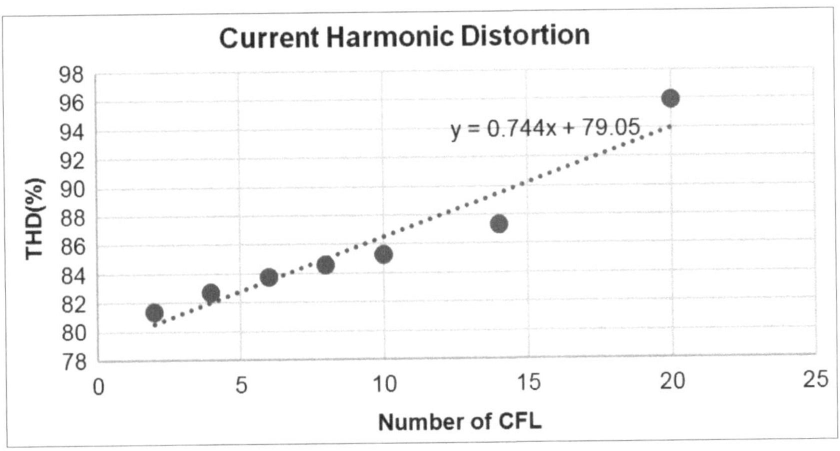

Fig. 16: THDi as Number of CFL Increase

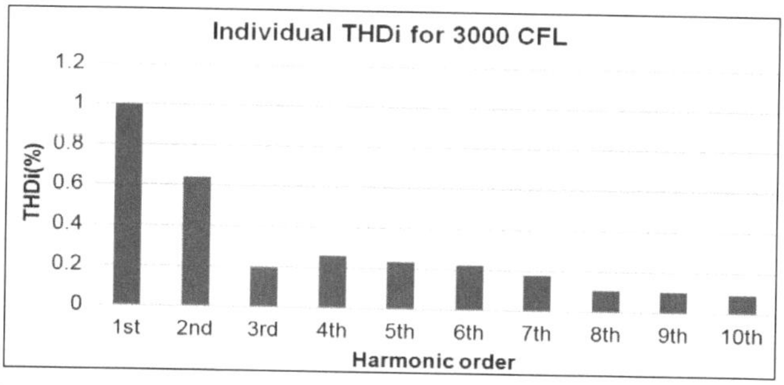

Fig. 17: Individual THDi for 3000 CFL *vs.* Harmonic Order

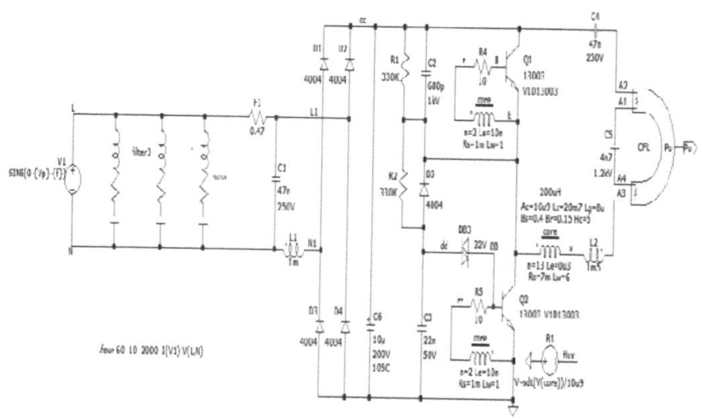

Fig. 18: CFL Circuit with Designed Shunt Passive Filter

Using above equation $Y = 0.7449X + 79.05$ where

Y is THDi

X is number of CFL

Then THDi if number of CFL is 3000

THDi = 0.7449 × 3000 + 79.05 = 2313.75%

Individual THDi for 3000 CFL is shown by Figure 17 as Bar graph.

As we observe from the Figure 17 as number CFLS increase the harmonic distortion that is created by them also increase and this harmonic will harm the components of power system. So, in order to reduce the damage that is caused by this harmonic and confront the THD with acceptable IEEE standard we develop the filter to mitigate harmonics.

CFL with Passive Filters Designed

The passive shunt filter networks were developed based on the calculated values of the inductances and resistances. Also, the capacitance of the capacitor (Cn) which was used to develop the filters was calculated to be 0.8 μF. The CFl circuit after designing shunt passive filter is shown in Figure 18.

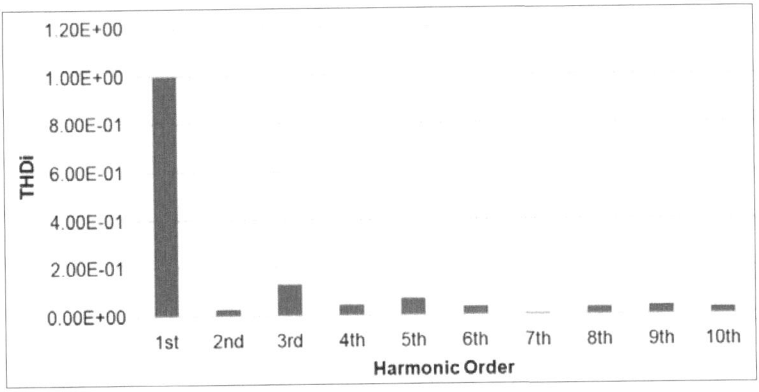

Fig 19: Current Total Harmonic Distortion vs. Harmonic Order for CFL with Filter

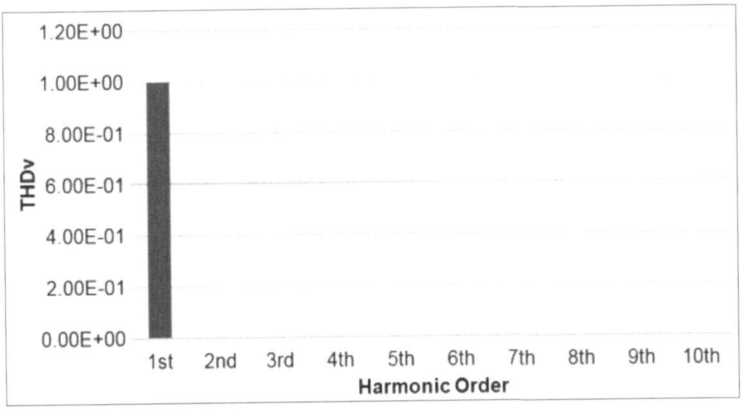

Fig. 20: Total Harmonic Distortion of Voltage vs. Harmonic Order for CFL with Filter

Results for the current and voltage total harmonic distortion of the FFT harmonics analysis after inserting filter were 17.14% and 0.000% respectively. Total harmonics order versus Fourier component was also drawn in Figures 19 and 20.

The Total Harmonic Distortion of voltage *vs.* Harmonic order is given as Figure 20.

Therefore, as we observed from the above discussion it is better to use filter in order to filter and mitigate the distortion that is caused by the switching devices that are found inside the circuit of CFL.

Conclusion

This research outlined a step-by-step process for modeling and simulating tiny fluorescent lamps for the purpose of analyzing and mitigating harmonic injection into power supply networks. Two methods were used for the harmonic analysis. These were the simplified FFT harmonics analysis algorithm and LTspice. The results obtained for THDi and THDv in relation to the fundamental current and voltage were 118.996% and 20.6006%. Simulink models of the CFL circuit with shunt passive harmonics filter networks were also developed with a view to evaluating their performance in terms of the CFL harmonics emission reduction. The choice of the method of harmonic reduction was based on its simplicity and cost. An FFT harmonics analysis was carried out on the developed Simulink models of the CFL circuits with the filters on LTspice software. The results obtained in relation to the fundamental current and voltage were 14.338% and 0.0009% for the THDi and THDv with the shunt filter. It was observed that the reductions in THD of the CFL were achieved with the shunt passive filters.

Due to its energy efficiency and other advantages, such as cost effectiveness and long life span, the compact fluorescent lamp has been the most widely used lighting device. The lamp is a nonlinear load. However, this type of lamp is associated with one major technical challenge, and that is the emission of harmonics. A harmonic is a multiple integer of the fundamental frequency. Even though CFL harmonic injections may be considered small, when a large number is used in a power distribution network, their effect will be a large harmonic source. Mitigation of these harmonics becomes very difficult once they are injected into the distribution network.

There has been much research on the analysis and mitigation of the injection of harmonics into the power distribution network. However, the methods used were characterized by inherent inaccuracy, complexity, and other challenges due to their large number of assumptions and high demands for computational effort. Consequently, this research work modeled and simulated compact fluorescent lamps for the analysis and mitigation of harmonics injection into power distribution networks and harmonic distortion caused by CFLs that are found around IOT student dormitories is also analyzed. As a future enhancement, the CFL prototype with series and shunt filters will be tested to determine its current and voltage total harmonic distortions. Furthermore, extend the research into other nonlinear loads, such as fans and PCs, with a view to mitigating the harmonics they generate.

References

[1] Afonso, *et al.* (2003). P-Q Theory power components calculations. In Industrial Electronics, 2003. ISIE'. 2003 *IEEE International Symposium on,* Vol. 1, pp. 385–390.

[2] Bollen and Gu (2006). Signal processing of power quality disturbances (Vol. 30). John Wiley & Sons.

[3] Collin and Meyer, J. (2014). Cancellation of harmonics between

group of modern compact fluorescent lamps. In Power Electronics, Electrical Drives, *Automation and Motion (SPEEDAM) International Symposium*, pp. 1190–1194.

[4] Cristaldi, L. and Salicone, S. (2002). A distributed system for electric power quality measurement IEEE measurement. *IEEE Transactions on Instrumentation and Measurement*, 51(4), 776–781.

[5] Cunill-Solà, J. and Salichs, M. (2007). Study and characterization of waveforms from low-watt (25 W) compact fluorescent lamps with electronic ballasts. *IEEE Transactions on power delivery*, 22(4): 2305–2311.

[6] Czarnecki, L.S. (2006). Instantaneous reactive power pq theory and power properties of three phase systems. *IEEE Transactions on Power Delivery*, 21(1), 362–367.

[7] Feng, C. *et al.* (2013). Tracing of energy consumption by using harmonic current. In: Research and Development (SCOReD), *2013 IEEE Student Conference* on (pp. 444–449).

[8] Franco, *et al.* (2015). Compact fluorescent lamps, LED lamps and harmonic distortion. *In Journal of Physics: Conference Series* (Vol. 575, No. 1, p. 012009). IOP Publishing.

[9] Franchetti, F. and Moura, J.M. (2009). Discrete Fourier transform on multicore. *IEEE Signal Processing Magazine*, 26(6), 90–102.

[10] Gufta, J.B. (2011). A Course in Electronics and Electrical Measurements and Instrumentation. 13th thoroughly revised edition. S.K. Kataria and sons, Daryaganj, Delhi, India, pp. 1002–1011.

[11] Gupta, S. *et al.* (2015). Power system harmonic analyzer based on FFT methods. *International Journal of Scientific Engineering and Applied Science (IJSEAS),* 1(3):199–206.

[12] Herrera, R.S. and S.P. (2007). Instantaneous reactive power theory: A comparative evaluation of different formulations. *IEEE transactions on power delivery*, 22(1), 595–604.

[13] Ingale, R. (2014). Harmonic Analysis Using FFT and STFT. *International Journal of Signal Processing, Image Processing and Pattern Recognition,* 7(4), 345–362.

[14] Igwilo, O.C. *et al.* (2013). *IOSR Journal of Electronics and Communication Engineering (IOSR–JECE),* 8(1):1–6.

[15] Ise, T., Hayashi, Y. and Tsuji, K. (2000). Definitions of power quality levels and the simplest approach for unbundled power quality services. In Harmonics and Quality of Power, 2000. *Proceedings. Ninth International Conference* on (Vol. 2, pp. 385–390). IEEE.

[16] Jabbar, R.A., *et al.* (2008). Impact of compact fluorescent lamp on power quality. In: Power Engineering Conference, 2008. AUPEC'08. Australasian Universities (pp. 1–5).

[17] John, G. Proakis and Dimitris, G. Manolakis (2007). Digital Signal Processing: Principles, Algorithms, and Applications. Asoke K. Ghosh, Prestice-Hill of India Private Limited, New Delhi, pp. 21–24.

[18] Khalil, M.R. and Ibrahem, A.T. (2013). Design and Implementation of FFT/IFFT System using Embedded Design Techniques. *International Journal of Engineering and Innovative Technology (IJEIT),* (3): 258–264.

Techno-Economic Analysis of HRES using Genetic Algorithm

Hemani Paliwal[1], Vikramaditya Dave[2] and Ankush Kohli

Department of Electrical Engineering,
College of Technology and Engineering,
MPUAT, Udaipur–313001, Rajasthan, India
E-mail: [1]hemani.8588@gmail.com; [2]vdaditya1000@gmail.com

ABSTRACT: *Renewable power generation can help countries to meet their sustainable development goals through provision of access to clean, secure, reliable and affordable energy. Scarcity of conventional energy resources, rise in fuel prices and harmful emissions from the burning of fossil fuels has made power generation from conventional energy sources unsustainable and unviable. In recent years, energy consumption has been significantly affected by globalization and the rapid pace of industrialization. The hybridization of renewable sources is a complex non-linear problem and the random availability of renewable sources makes the designing process more complicated. In this study, a grid-connected solar/wind/biogas HRES is proposed to fulfill the load demand of the Electrical Engineering department of the College of Technology and Engineering situated at Udaipur. To evaluate the techno-economic feasibility of the system, optimal cost, size, and emission calculations of the system, an artificial intelligence technique namely Genetic Algorithm (GA) is proposed. The proposed technique is found effective and suitable in designing HRES; gives minimum cost, optimal size, and reduced emissions at high reliability.*

Keywords: Optimization, Solar, Wind, Biogas, Genetic Algorithm.

Introduction

Renewable energy sources are gradually touching everyone's life in different ways. Renewable energy sources are not only an alternative sources but an exotic option to move towards the clean and green energy. Renewable energy can bring not only carbon free energy but also provide affordable electricity to all, social equality and sustainable development. Exponential growth in population, increasing energy demand, draining of fossil fuels and environmental concerns, led to strides toward alternative sources of energy [1]. Looking towards drastic climatic changes, India has moved forward to use renewable energy sources for electricity generation as the country has plenty of renewable energy sources. With a population of 1.3 billion, India is the second most populous country in the world and the third-largest economy, measured by purchasing power parity (PPP). India has seen strong economic performance in recent decades, enabling a significant decrease in poverty levels, greater energy access for its citizens and growing penetration of cleaner energy across the economy. India has announced world's biggest programmes in all states for renewable energy development [2-5].

Energy sector in India is progressing towards an enduring development. Over the period 2005–2014, India has reduced its emission intensity approximately 20% by taking bold actions and continuing efforts [6-7].

The estimated solar power in India during 1995 was 5 × 1015 KWh/pa, wind power was 20 GW, Bioenergy was 17 GW, bagasse was 8 GW and small hydropower was 10 GW. In 2006, estimated renewable energy in India was 85 GW, with 35 MW solar energy, 4500 MW wind energy, biomass 25 GW and small hydro power of 15 GW. In 2018, estimated renewable energy was 1096.080 GW, according to MNRE report 2017–18. There is a growth of approximately 6 GW of renewable energy capacity in the year 2019–20 (Energy Statistics 2020) [8-10].

Hybrid Renewable Energy Systems

Renewable energy sources are environmentally clean without causing greenhouse gas and cheap alternative to fossil fuels. To overcome or at least limit some of the problems associated conventional energy sources, renewable energy resources are the solution [11].

Renewable energy resources are highly site-specific, stochastic in nature and are fairly evenly distributed around the world with little or no costs. They are greatly dependent on the climatic conditions, geographical factors and seasons of the site under consideration [12].

Hybrid renewable energy systems uses the combination of two or more renewable sources or renewable sources with conventional sources and storage devices such as wind and PV system or wind and hydro system or wind PV system with battery or wind PV system with diesel generator for round the clock generation of electrical power. It is commonly used as standalone mode for remote or islanded areas to utilize locally available resources to fulfil load demand of the area. This reduces transmission and distribution losses and hence reduces cost [13-15].

Genetic Algorithms (GA)

Genetic Algorithm (GA) is a search-based optimization technique based on the principles of Genetics and Natural Selection. Nature has always been a great source of inspiration to all mankind. GAs is a subset of a much larger branch of computation known as evolutionary computation. It is frequently used to find optimal or near-optimal solutions to difficult problems which otherwise would take a lifetime to solve. It is frequently used to solve optimization problems, in research, and in machine learning [16-19].

GAs was developed by John Holland and his students and colleagues at the University of Michigan, most notably David E. Goldberg and has since been tried on various optimization problems with a high degree of success. Genetic algorithm is an optimization technique used to find solutions of complex optimization problems with or without constrained parameters. It is a heuristic approach of global search. GA can be applied to any branch of engineering and sciences to solve optimization problem. GA is suitable for complex problems where traditional hill climbing methods fail [20-22].

Optimization Problem

Optimization of size, economic operation, controller design, real and reactive power control, voltage and frequency, reliability control are the main issues of optimization problem in the area of HRES [23]. The optimization problem include the knowledge of availability of renewable resources for power generation, hybrid system design, i.e. sizing and storage system options, interconnection of different sources with the help of converter, load curve of the area that has to be met, optimal operation, optimized cost, battery life, voltage and frequency regulation in case of grid connected system and overall effect on environment. Numerous researches have been made for optimum economic design of HRES using intelligent techniques and algorithm. Economic designing of HRES means finding the best combination of renewable or conventional generating sources (size) with or without battery, which can securely fulfill the load demand [24-25]. A detailed analysis on the most popular algorithms available from literature survey, namely Genetic Algorithms (GA), Particle Swarm Optimization Algorithm (PSO) and Fuzzy logic (FL), along with discussions on new optimization algorithms which have been applied for the control and optimization of the hybrid renewable energy based system is contributed in the following subsections [26-27].

Various soft computing and optimization based approaches are gaining popularity in the solution of optimization problems with multiple conflicting objectives in recent times. However from the literature survey it has been observed that very few research works have been found wherein the focus has been on making a comparative assessment of the use and application of such soft computing and optimization algorithms (Chandel *et al.*, 2014, Rajalaxshmi *et al.* 2020) [28-29].

Objective Function

An objective function is a mathematical function in terms of the identified decision variables which expresses the

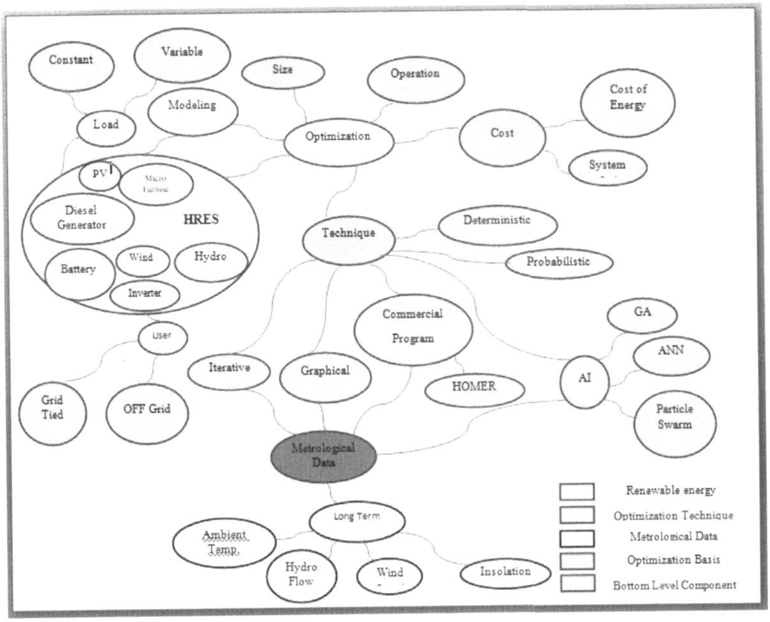

Fig. 1: Optimization Problem

target to be reached in an optimization problem [30]. In this optimal design study, the aim is to minimize the total cost of Energy and carbon emissions of the proposed HRES. The optimization parameters are number of PV arrays (NPv), wind turbines (NwT), size of biogas generator (Nbio) and size of inverter (Ninv). All of these parameters are optimized within a specific range between maximum and minimum value. The total annualized cost (TAC) of the system is considered, so the objective function is defined by:

Minimize *(COE(x), Emission CO_2(x))*

Where

$$COE(₹/KWh) = \frac{TAC}{E_{served}} \quad \dots (1)$$

Total annual cost is given by:

$$TAC = \sum C_{cap}\,(PV/WT/Bio) + C_{O\&M}\,(PV/WT/Bio) + C_{Rep}\,(PV/WT/Bio) + FC\,(bio) - SV\,(PV/WT/Bio) + \text{Grid cost} \quad \dots (2)$$

Where

C_{cap} – Capital Cost

$C_{O\&M}$ – Operation and maintenance cost

C_{Rep} – Replacement Cost

FC – Fuel cost

SV – Salvage cost

$$C_{cap} = (P_{PV} \times C_{PV} + P_{WT} \times C_{WT} + P_{bio} \times C_{bio} + P_{con} \times C_{con}) \times CRF \quad \dots (3)$$

Where

$$CRF\,(i,T) = \frac{i(1+i)^T}{(1+i)^T - 1} \quad \dots (4)$$

The value of the equivalent annuity of a present worth is obtained by multiplying the present worth by the capital recovery factor (CRF). Here i denote the interest rate, and T is the lifespan of the system [31]. In this study, the interest rate is set at 6% per year and the lifespan of the system at 20 years.

$$\text{Solar Cost} = N_{PV} \times [C_{O\&M}^{PV} + C_{cap}^{PV} \times CRF] \quad \dots (5)$$

$$\text{Wind Cost} = N_{WT} \times [C_{O\&M}^{WT} + C_{cap}^{WT} \times CRF] \quad \dots (6)$$

$$\text{Biogas Cost} = N_{bio} \times [C_{O\&M}^{bio} + (C_{cap}^{bio} + C_{rep}^{bio}) \times CRF \quad \dots (7)$$

$$\text{Grid cost} = \text{Grid purchase} \times \text{puchase rate} - \text{Grid sale} \times \text{sale rate} \quad \dots (8)$$

$$\text{Inverter cost} = N_{inv} \times [C_{O\&M}^{inv} + C_{cap}^{inv} \times CRF] \quad \dots (9)$$

Where N-No. of Components (Solar PV, Wind Turbine, Inverter, Biogas Generator)

$$\textbf{Total Annual cost} = \text{Solar cost} + \text{wind cost} + \text{biogas cost} + \text{grid cost} + \text{inverter cost} \quad \dots (11)$$

Net Present Cost (NPC): The net present cost is the investment costs plus the discounted present values of all future costs during the lifetime of the system [32],

$$NPC = \frac{TAC}{CRF} \quad \dots (12)$$

Modeling and Simulation

The site selected for the study of multi-objective optimization of Hybrid Renewable energy sources (HRES) is Electrical Engineering Department of College of Technology and Engineering, MPUAT, Udaipur Rajasthan. The location of the site is located between lines of Latitudes 24 degrees 35.13 minutes North and lines of Longitudes 73 degrees 42.75 minutes East in Udaipur district of Rajasthan.

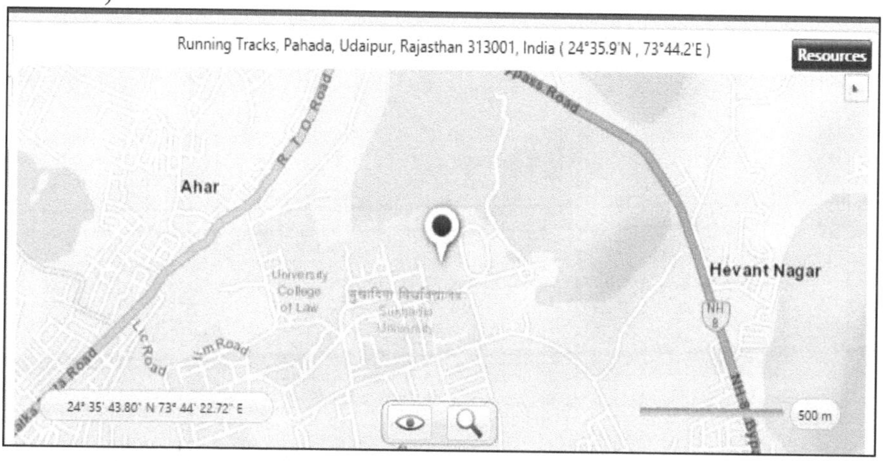

Fig. 2: Google Earth Image of Electrical Engineering Department

Load Data

The load demand of the Electrical Engineering Department mainly constituted as lighting load, AC load, computer and printers load, equipments of laboratory and machines etc. The approximate connected load of the department is calculated through an energy audit survey of the department and the duration of load utilization is mainly during 9 a.m. to 6 p.m. and remaining hours have only basic lighting load of the department. Weekdays have load from 9 a.m. to 6 p.m. while weekends have only basic lighting load. From the survey of the department, the list of electrical devices and equipments, and their ratings are collected. With the help of collected data (for 8760 hours of the year), study of past year electricity bills and rigorous review of literature, the load profile of the department is recorded to estimate the actual consumption of the department.

Meteorological Data

Meteorological data consists of physical parameters that are measured directly by instrumentation, and include ambient temperature, wind speed, wind directions, sky clearness index and solar insolations [33-34]. There is huge potential availability of renewable resources at the site selected i.e. Electrical Engineering Department of CTAE.

Solar Resources

Monthly average solar global horizontal irradiance data are taken from National Renewable Energy Laboratory (NREL) database and hourly solar radiations are taken from PVGIS-SARAH [35]. The maximum solar radiations are found in the May month, 7.199 KWh/m^2/day whereas minimum solar radiations are recorded in month of December i.e. 4.307 KWh/m^2/day as shown in Table 4.3. The average annual scaled solar radiations are 5.63 KWh/m^2/day. Monthly average solar radiations are shown in Figure 4.

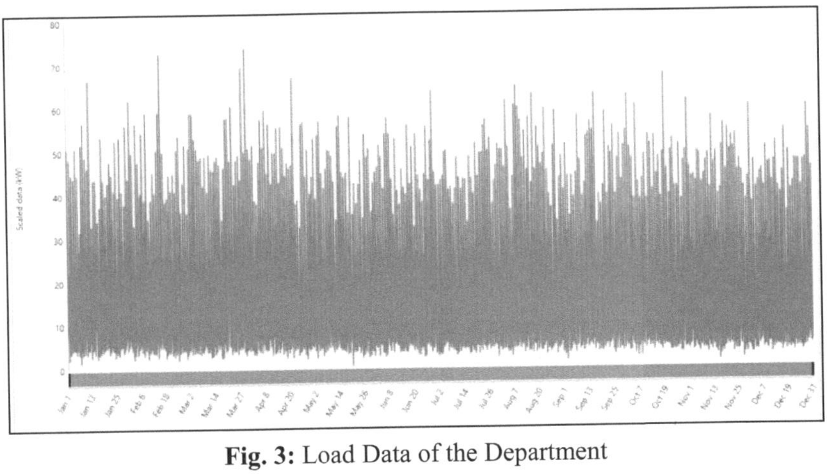

Fig. 3: Load Data of the Department

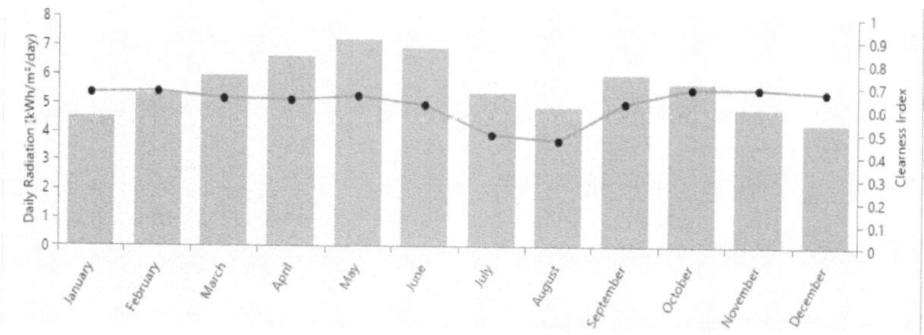

Fig. 4: Monthly Solar Radiations

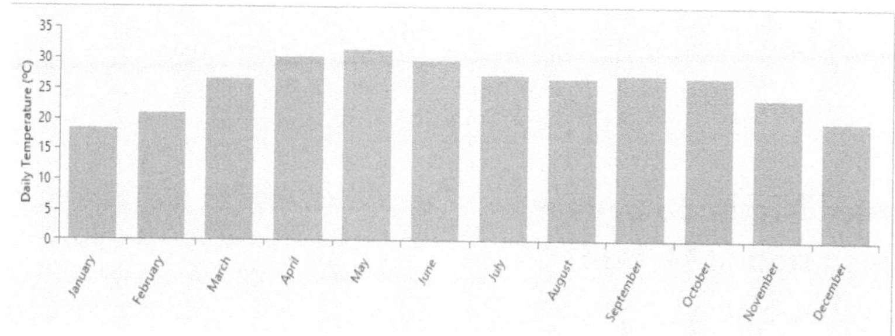

Fig. 5: Ambient Temperature

Ambient Temperature

As ambient temperature of the site is important for solar power calculations, as shown in equation 3.9, the hourly ambient temperature of the location are recorded from PVGIS-SARAH and the monthly average ambient temperature in °C are shown in Fig. 5. Temperature of location have significant effect on wind speeds too, therefore hourly wind speeds are considered in this study [36].

Wind Resources

Wind resources are determined using NASA surface methodology and wind atlas where wind direction is considered at 50 meters above the earth surface [37-38]. The hourly wind speed is provided as the database for the given month over the complete year. The average annual wind speed is a good indicator for running a wind turbine at a given location. The highest wind speeds are observed in the month of May which is 5.6 m/s and lowest wind speed is 3.52 m/s in the month of October. The annual average wind speed for the location is 4.43 m/s.

Biomass Resources

On the basis of availability of feedstock at the location, monthly average biomass (tones/day) is considered [39]. The LHV of biogas is 5.5 MJ/Kg, carbon content is 5% and gasification ratio is 0.7 Kg/Kg. Average requirement of biogas raw material is 0.33 Tonne/day.

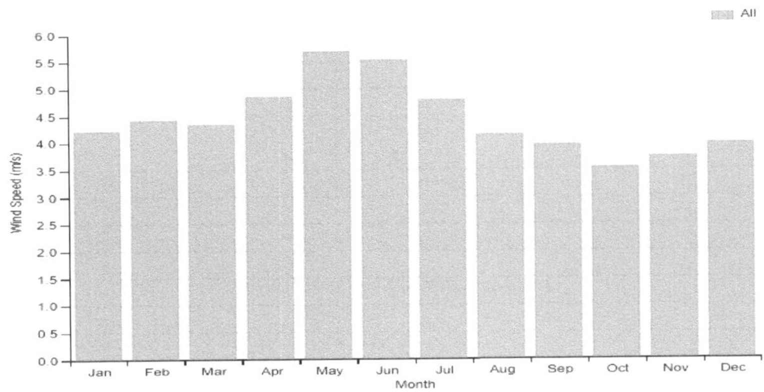

Fig. 6: Monthly Wind Speeds

Proposed Methodology for Optimization of Hybrid Renewable Energy Sources

Majority of the research scientists developed various configurations of hybrid renewable energy system models [40]. From the available literature and gaps identified in the research recognized above, here a novel hybrid renewable energy system (HRES) model is developed for modeling and optimization of a grid connected solar wind and bio-gasifier HRES for electrification in Electrical Engineering Department of College of Technology

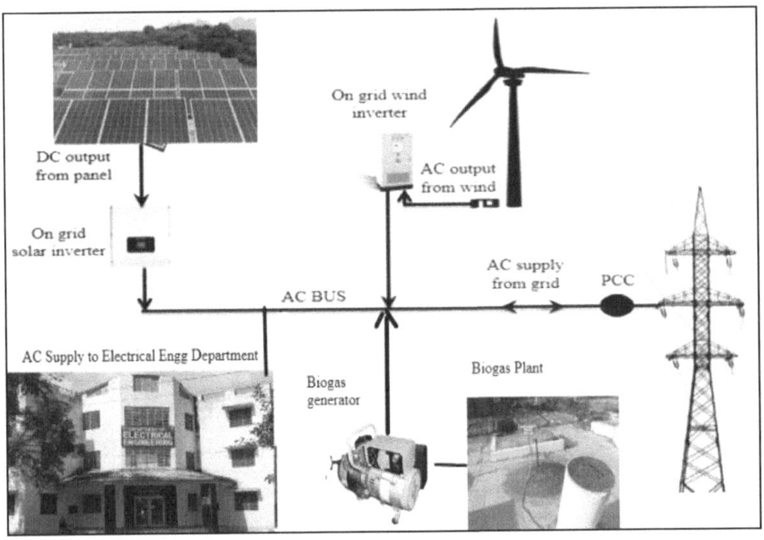

Fig. 7: Proposed Schematic Diagram

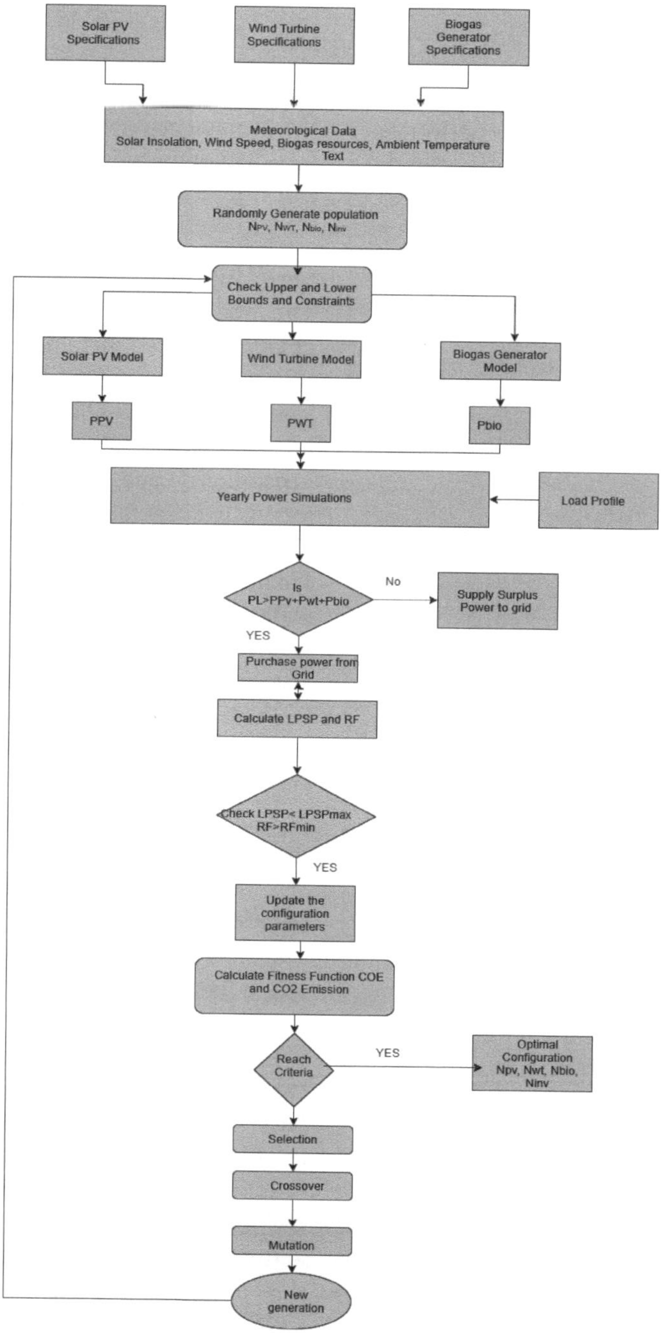

Fig. 8: Flowchart of Procedure of Genetic Algorithm

and Engineering, MPUAT, Udaipur. Multi-objective Genetic Algorithm is proposed to solve sizing and optimization problems. System performance will be evaluated and compared in terms of minimum value of NPC, COE and minimum CO_2 emission.

To minimize the fitness function using the ga function, a function (economic_analysis.m file) handle the fitness function as well as specifying the number of variables (nvar=3) as the second argument. Lower and upper bounds are provided as LB and UB respectively.

The GA solver handles linear constraints and bounds differently from nonlinear constraints. All the linear constraints and bounds are satisfied throughout the optimization. However, GA may not satisfy all the nonlinear constraints at every generation. If GA converges to a solution, the nonlinear constraints will be satisfied at that solution [41]. GA uses the mutation and crossover functions to produce new individuals at every generation. The way the GA satisfies the linear and bound constraints is to use mutation and crossover functions that only generate feasible points. All the crossover functions in the toolbox generate points that satisfy the linear constraints and bounds [42].

The maximum population size in the specified GA program is 600, the stochastic uniform function is used for selection, the Heuristic crossover function (cross-over probability 85%) for cross-over operation and the Gaussian mutation function (with mutation probability rate 3%) for mutation operation are used. The optimization process terminated when the average change in the fitness value less than options or constraint violation is less than options [43].

Results

The optimization program starts calculating fitness function and calculates best and means value of that function. The maximum numbers of generations are 600, but the optimizer runs up to 92 generations and the 'Optimization terminated' as the average change in the fitness value less than options. The Pareto-optimal fronts for fitness value and generations of grid connected solar-wind-biogas hybrid system using the proposed approach is shown in Figure 9. From Figure 9 one can observe that the non-dominated solutions are smoothly.

The final result provided by this iterative technique is shown in Table 1. The optimal design according to GA output consists of 52.36 KW solar PV systems, 10 KW wind turbine system and a biogas genset of 20 KW. The proposed system is used in grid connected mode, energy balance calculations are performed on hourly basis (8760 hours). Whenever the calculated power by renewable sources is found less than the load demand of the location grid purchases are made and when the load demand is less than the power generated by renewable sources, the surplus power is sold to utility grid.

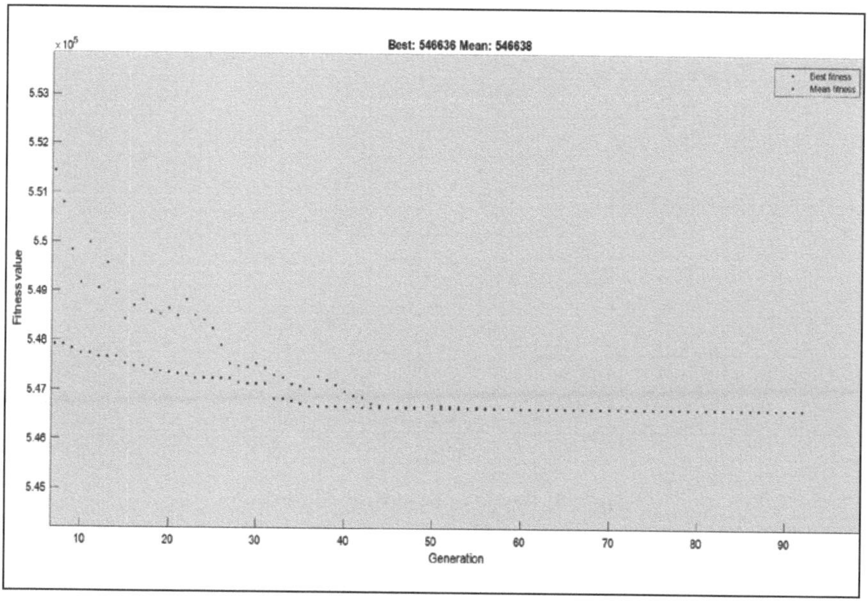

Fig. 9: Pareto Front of GA

Table 1: Optimal Solution by GA Output

Size	Solar PV system	52.36 KW
	Wind turbine	10 KW
	Biogas genset	20 KW
Energy Generated	Solar PV system	92,270 KWh
	Wind turbine	5585.6 KWh
	Biogas genset	41,505 KWh
Annualized cost	Solar PV system	₹ 1,75,320
	Wind turbine	₹ 64,136
	Biogas genset	₹ 2,28,740
	Net grid cost	₹ 1,15,620
	Inverter cost	₹ 20,552
Grid Exchange	Grid purchase	21,260 KWh
	Grid Sales	38,706 KWh
Annual Cost		₹ 5,86,640
Net Present Cost (NPC)		₹ 78,64,500
Cost of Energy (COE)		₹ 3.5563

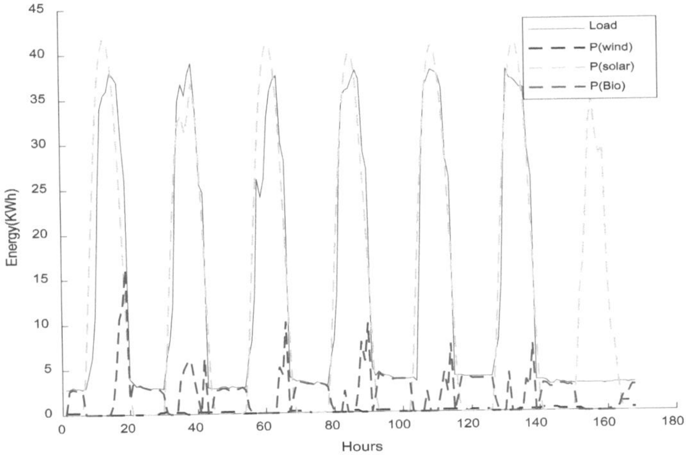

Fig. 10: Results of Genetic Algorithm

Discussion

Interest in generating electricity from renewable energy is increasing world-wide. With the very deep reliance on fossil fuels, the following details are frightening: The coal and oil manufacture in the world will achieve its greatest threshold soon, thereafter, it will keep on decreasing. World uranium production is expected to attain its zenith level in 2035. Nuclear fuel can be recycled, but it is a riskier procedure and the clearance of the spent fuel is also at high risk. So the total world is giving consideration towards the Renewable Energy System (RES). To further address the environmental issues arising from the obnoxious emissions, more efficient supercritical coal-based units are being commissioned and old and inefficient coal-based capacity is being retired. Renewable energy sources are environmentally clean without causing greenhouse gas and cheap alternative to fossil fuels. To overcome or at least limit some of the problems associated with conventional energy sources, renewable energy resources are the solution. Renewable energy resources are highly site-specific, stochastic in nature, and are fairly evenly distributed around the world with little or no costs. They are greatly dependent on the climatic conditions, geographical factors, and seasons of the site under consideration. Therefore Hybrid Renewable Energy Systems (HRES) having integration of fossil fuels with renewable sources is the best solution. However, for better performance of HRES, environmental factors and a proper combination of suitable renewable energy sources is very important. System reliability during various climatic conditions and the overall cost of the system is a serious concern in designing HRES. Therefore grid-connected HRES is the best option for urban areas where a utility grid is available but using HRES can reduce dependency on the grid and reduces emission in the atmosphere.

HRES has a large number of design options that should be wisely chosen and uncertainty in its input parameters makes the system unpredictable. So the overall objective of this study is to design an optimal HRES which is cost economic and reliable throughout the year. Here a grid-connected solar-wind-biogas system is proposed to fulfill the energy demands of the Electrical Engineering Department of CTAE, Udaipur Rajasthan. For solving the non-linear optimization problems, the HOMER Pro software tool is used. The result obtained from HOMER pro software gives optimal size and cost of the system. It proposed a 58.2 KW solar PV system, 10 KW wind turbine, and 18 KW Biogas Genset connected to the grid. The Net Present Cost (NPC) of the system is ₹81,21,959.22 and the cost of energy (COE) of this configuration is ₹3.77.

Then the optimal system is further validated with the artificial intelligent technique, Genetic algorithm. The multi-objective optimization of HRES is done on MATLAB based GA program. The results of the GA program are also well simulated and fast converging. The optimization program terminates after iterations up to 92 generations and gives the optimal size of the system with reduced NPC and COE compared to HOMER Pro software. The optimal design according to GA output consists of a 52.36 KW solar PV system, 10 KW wind turbine system, and a biogas Genset of 20 KW. The Net Present Cost (NPC) of the system is ₹ 78,64,500 and the cost of energy (COE) of this configuration is ₹ 3.5563.

The comparison of results shows that the GA program converges faster than the HOMER optimizer tool and provides the optimal design of the system that has lower sizing options, COE and NPC compared to HOMER. Also, emission calculation of both the techniques gives a considerable saving of CO_2 per year with the proposed HRES.

With the study of software tools and artificial intelligence techniques, it can be concluded that results obtained from the GA are superior compared to the HOMER tool. This is because the HOMER tool performs simulation and optimizations by simply energy balance calculations of all available options whereas GA uses biologically inspired operators like selection, crossover, and mutation to produce better output in each iteration.

References

[1] Saiprasad, N., Kalam, A. and Zayegh, A., "Zero-emission renewable energy system model for a community in India," *Asia-Pacific Power Energy Eng. Conf. APPEEC,* Vol. 2017-Nov., pp. 1–6, 2018, doi: 10.1109/APPEEC.2017.8308942.

[2] Dawoud, S.M., Lin, X. and Okba, M.I., "Hybrid renewable microgrid optimization techniques: A review," *Renew. Sustain. Energy Rev.*, Vol. 82, pp. 2039–2052, Feb. 2018, doi: 10.1016/j.rser.2017.08.007.

[3] Deshmukh, M.K. and Deshmukh, S.S., "Modeling of hybrid renewable energy systems," Renew. Sustain. Energy Rev., Vol. 12, No. 1, pp. 235–

249, 2008, doi: 10.1016/j.rser.2006.07.011.

[4] Rout, K. and Sahu, J.K., "Various Optimization Techniques of Hybrid Renewable Energy Systems for Power Generation : A Review," *Int. Res. J. Eng. Technol.,* Vol. 5, No. July, pp. 1173–1176, 2018.

[5] Sambaiah, K.S., "A review on optimal allocation and sizing techniques for DG in distribution systems," *Int. J. Renew. Energy Res.,* Vol. 8, No. 3, pp. 1236–1256, 2018.

[6] Tezer, T., Yaman, R. and Yaman, G., "Evaluation of approaches used for optimization of stand-alone hybrid renewable energy systems," *Renew. Sustain. Energy Rev*., Vol. 73, No. January, pp. 840–853, Jun. 2017, doi: 10.1016/j.rser.2017.01.118.

[7] Sawle, Y., Gupta, S.C. and Bohre, A.K., "Review of hybrid renewable energy systems with comparative analysis of off-grid hybrid system," *Renewable and Sustainable Energy Reviews,* Vol. 81. Elsevier Ltd, pp. 2217–2235, 01-Jan-2018, doi: 10.1016/j.rser.2017.06.033.

[8] Sinha, S. and Chandel, S.S., "Review of software tools for hybrid renewable energy systems," *Renewable and Sustainable Energy Reviews,* Vol. 32. pp. 192–205, Apr-2014, doi: 10.1016/j.rser.2014.01.035.

[9] Singh, A., Baredar, P. and Gupta, B., "Computational Simulation & Optimization of a Solar, Fuel Cell and Biomass Hybrid Energy System Using HOMER Pro Software," *Procedia Eng.,* Vol. 127, pp. 743–750, 2015, doi: 10.1016/j.proeng.2015.11.408.

[10] Saharia, B. Jyoti; Brahma, H. and Sarmah, N., "A review of algorithms for control and optimization for energy management of hybrid renewable energy systems," *J. Renew. Sustain. Energy,* Vol. 10, No. 5, 2018, doi: 10.1063/1.5032146.

[11] Siddaiah, R. and Saini, R.P., "A review on planning, configurations, modeling and optimization techniques of hybrid renewable energy systems for off grid applications," *Renewable and Sustainable Energy Reviews,* Vol. 58. 2016, doi: 10.1016/j.rser.2015.12.281.

[12] Khare, V., Nema, S. and Baredar, P., "Optimisation of the hybrid renewable energy system by HOMER, PSO and CPSO for the study area," *Int. J. Sustain. Energy,* Vol. 36, No. 4, 2017, doi: 10.1080/14786451.2015.1017500.

[13] Singh, R., Bansal, R.C., Singh, A.R. and Naidoo, R., "Multi-Objective Optimization of Hybrid Renewable Energy System Using Reformed Electric System Cascade Analysis for Islanding and Grid Connected Modes of Operation," *IEEE Access*, Vol. 6, No. August, pp. 47332–47354, 2018, doi: 10.1109/ACCESS.2018.2867276.

[14] Twaha, S. and Ramli, M.A.M., "A review of optimization approaches for hybrid distributed energy generation systems: Off-grid and grid-connected systems," *Sustain. Cities Soc.,* Vol. 41, pp. 320–331, Aug. 2018, doi: 10.1016/j.scs.2018.05.027.

[15] Benhachani, Z., Azoui, B., Abdessemed, R. and Chabane, M., "Optimal sizing of a solar-wind hybrid system supplying a farm in a semi-arid region of Algeria," *Proc. Univ. Power Eng. Conf.,* pp. 1–6, 2012, doi: 10.1109/UPEC.2012.6398688.

[16] Akram, U., Khalid, M. and Shafiq, S., "Optimal sizing of a wind/solar/ battery hybrid grid-connected micro-grid system," *IET Renew. Power Gener.,* Vol. 12, No. 1, pp. 72–80, 2018, doi: 10.1049/iet-rpg.2017.0010.

[17] Alalwan, S.H. and Kimball, J.W., "Optimal sizing of a wind/solar/ battery hybrid microgrid system using the forever power method," *IEEE Green Technol. Conf.,* Vol. 2015-July, pp. 29–35, 2015, doi: 10.1109/GREENTECH.2015.21.

[18] Nacer, T., Nadjemi, O. and Hamidat, A., "Optimal sizing method for grid connected renewable energy system under Algerian climate," *in IREC2015 the Sixth International Renewable Energy Congress*, 2015, pp. 1–5, doi: 10.1109/IREC.2015.7110956.

[19] Kumar, A.P., "Analysis of Hybrid Systems: Software tools," *Proceeding IEEE - 2nd Int. Conf. Adv. Electr. Electron. Information, Commun. Bio-Informatics, IEEE - AEEICB 2016,* pp. 327–330, 2016, doi: 10.1109/AEEICB.2016.7538302.

[20] Tank, V., Bhutka, J. and Harinarayana, T. (2016) Wind Energy Generation and Assessment of Resources in India. *Journal of Power and Energy Engineering,* 4, 25-38. http://dx.doi.org/10.4236/jpee.2016.410002

[21] Rawat, R. and Chandel, S.S., "Simulation and optimization of solar photovoltaic-wind stand alone hybrid system in hilly terrain of India," *Int. J. Renew. Energy Res.,* Vol. 3, No. 3, pp. 595–604, 2013, doi: 10.20508/ijrer.88774.

[22] Salam, M.A., Aziz, A., Alwaeli, A.H.A. and Kazem, H.A., "Optimal sizing of photovoltaic systems using HOMER for Sohar, Oman," *Int. J. Renew. Energy Res.,* Vol. 3, No. 3, pp. 470–475, 2013, doi: 10.20508/ijrer.57520.

[23] Alkarrami, F., Iqbal, T. and Pope, K., "Optimal sizing of a stand-alone hybrid energy system for water pumping in Sirte, Libya," *2016 IEEE Electr. Power Energy Conf. EPEC 2016,* pp. 1–5, 2016, doi: 10.1109/EPEC.2016.7771679.

[24] Ahmad, F. and Alam, M.S., "Optimal Sizing and Analysis of Solar PV, Wind, and Energy Storage Hybrid System for Campus Microgrid," *Smart Sci.,* Vol. 6, No. 2, pp. 150–157, Apr. 2018, doi: 10.1080/23080477.2017.1417005.

[25] Singh, A. and Baredar, P., "Power sharing and cost optimization of hybrid renewable energy system for academic research building," *J. Electr. Eng. Technol.,* Vol. 12, No. 4, pp. 1511–1518, 2017, doi: 10.5370/JEET.2017.12.4.1511.

[26] Navamani, J. Divya, Lavanya, A., Prahadheeshwar, C.M. and Riyazudeen, S. Mohammed, "Hybrid power system design using homer pro," *Int. J. Recent Technol. Eng.,* Vol. 8, No. 1 Special Issue4, pp. 605–609, 2019.

[27] Dehwah, A.H.A. and Krarti, M., "Optimal Hybrid Power Energy Systems for Residential Communities in Saudi Arabia," *J. Sol. Energy Eng.*, Vol. 141, No. 6, pp. 1–10, Dec. 2019, doi: 10.1115/1.4043633.

[28] Suresh, V., Muralidhar, M. and Kiranmayi, R., "Modelling and optimization of an off-grid hybrid renewable energy system for electrification in a rural areas," *Energy Reports,* Vol. 6, pp. 594–604, 2020, doi: 10.1016/j.egyr.2020.01.013.

[29] Charabi, Y. and Abdul-Wahab, S., "Wind turbine performance analysis for energy cost minimization," *Renewables Wind. Water, Sol.,* Vol. 7, No. 1, 2020, doi: 10.1186/s40807-020-00062-7.

[30] Merlin, V.L., Santos, R.C. dos, Pavani, A.P. Grilo, Coury, D.V., Oleskovicz, M. and Vieira, J.C. de Melo, "Artificial Neural Network Based Approach for Anti-islanding Protection of Distributed Generators," *J. Control. Autom. Electr. Syst.,* Vol. 25, No. 3, pp. 339–348, Jun. 2014, doi: 10.1007/s40313-013-0096-0.

[31] Baygi, S.M.H., Elahi, A. and Karsaz, A., "A novel framework for optimal sizing of hybrid stand-alone renewable energy system: A gray Wolf optimizer," *3rd Conf. Swarm Intell. EVol. Comput. CSIEC 2018,* pp. 1–7, 2018, doi: 10.1109/CSIEC.2018.8405415.

[32] Farahmand, M.Z., Nazari, M.E. and Shamlou, S., "Optimal sizing of an autonomous hybrid PV-wind system considering battery and diesel generator," *2017 25th Iran. Conf. Electr. Eng. ICEE 2017,* No. 8, pp. 1048–1053, 2017, doi: 10.1109/IranianCEE.2017.7985194.

[33] Okunade, P., Ansari, M., Asrari, A. and Khazaei, J., "Application of Optimization for Daily Scheduling of Renewable Distributed Generations Considering Market Profits in Distribution Networks," *2018 North Am. Power Symp. NAPS 2018*, pp. 1–6, 2019, doi: 10.1109/NAPS.2018.8600654.

[34] Rajanna, S. and Saini, R.P., "Development of optimal integrated renewable energy model with battery storage for a remote Indian area," *Energy,* Vol. 111, pp. 803–817, Sep. 2016, doi: 10.1016/j.energy.2016.06.005.

[35] Abdelshafy, A.M., Hassan, H., Mohamed, A.M., El-Saady, G. and Ookawara, S., "Optimal grid connected hybrid energy system for Egyptian residential area," *Proceeding - ICSEEA 2017 Int. Conf. Sustain. Energy Eng. Appl. " Continuous Improv. Sustain. Energy Eco-Mobility",* Vol. 2018-Janua, No. October, pp. 52–60, 2017, doi: 10.1109/ICSEEA.2017.8267687.

[36] Soni, V.K. and Khare, R., "Optimal sizing of HRES for small sized institute using HOMER," *Proc. 2014 IEEE 2nd Int. Conf. Electr. Energy Syst. ICEES 2014,* pp. 77–81, 2014, doi: 10.1109/ICEES.2014.6924145.

[37] Gurubel, K.J., Osuna-Enciso, V., Cardenas, J.J., Coronado-Mendoza, A., Perez-Cisneros, M.A. and Sanchez, E.N., "Neural forecasting and optimal sizing for hybrid renewable energy systems with grid-connected storage system," *J. Renew. Sustain. Energy,* Vol. 8, No. 4, 2016, doi: 10.1063/1.4960125.

[38] Mokhtara, C., Negrou, B., Settou, N., Settou, B. and Samy, M.M., "Design optimization of off-grid Hybrid Renewable Energy Systems considering the effects of building energy performance and climate change: Case study of Algeria," *Energy,* Vol. 219, p. 119605, 2021, doi: 10.1016/j.energy.2020.119605.

[39] Golestani, S. and Tadayon, M., "Distributed generation dispatch optimization by artificial neural network trained by particle swarm optimization algorithm," *2011 8th Int. Conf. Eur. Energy Mark. EEM 11,* No. May, pp. 543–548, 2011, doi: 10.1109/EEM.2011.5953071.

[40] Amer, M., Namaane, A. and M'Sirdi, N.K., "Optimization of Hybrid Renewable Energy Systems (HRES) Using PSO for Cost Reduction," *Energy Procedia*, Vol. 42, pp. 318–327, 2013, doi: 10.1016/j.egypro.2013.11.032.

[41] Khenfous, S., Kaabeche, A., Bakelli, Y. and Sba, K.M., "Optimal size of renewable hybrid system applying nature-inspired algorithms," *2018 Int. Conf. Wind Energy Appl. Alger. ICWEAA 2018*, No. 1, pp. 1–6, 2019, doi: 10.1109/ICWEAA.2018.8605047.

[42] Sanajaoba, S. and Fernandez, E., "Maiden application of Cuckoo Search algorithm for optimal sizing of a remote hybrid renewable energy System," *Renew. Energy,* Vol. 96, pp. 1–10, 2016, doi: 10.1016/j.renene.2016.04.069.

[43] Nadjemi, O., Nacer, T., Hamidat, A. and Salhi, H., "Optimal hybrid PV/wind energy system sizing: Application of cuckoo search algorithm for Algerian dairy farms," *Renew. Sustain. Energy Rev.*, Vol. 70, No. November, pp. 1352–1365, 2017, doi: 10.1016/j.rser.2016.12.038.

Equation based Modeling of Photovoltaic System in Matlab/Simulink

Narendra Sankhla[1], Khamma Kanwar[2] and Santosh Kumari Meena[3]

Department of Electrical Engineering, M.B.M. Engineering College, Jodhpur–342011, Rajasthan, India

E-mail: [1]narendrasankhla1234@gmail.com; [2]kkjoya.jnvu@gmail.com

ABSTRACT: *In this paper a DS-100M Solar Photovoltaic Panel is selected for modeling. The mathematical model of PV system has been developed using manufacturer datasheet parameters and mathematical equations in MATLAB/ Simulink software. Then I-V and PV characteristics have been plotted. The electrical characteristics of the panel provide information about the operational state of the system.*

Keywords: Photovoltaic Panel, MATLAB/Simulink.

Introduction

Solar photovoltaic (PV) systems have become the fast-growing alternative source of energy. These systems are environmentally friendly and require less maintenance. The output power from a PV system depends on the surrounding irradiance and temperature [1].

A PV cell is a p-n junction fabricated in a thin wafer of semiconductors. The electro-magnetic radiation of solar energy can be directly converted to electricity through the photovoltaic effect. The photons with energy greater than the band-gap energy of the semiconductor create some electron-hole pairs proportional to the incident irradiation. They start moves between the top and bottom surfaces of the semi-conducting material. This movement of electrons generates a current known as a direct current (DC). Then it can be converted into alternating current (AC) using an inverter for use in the home.[2]

The PV-based systems are employed in stand-alone applications such as street lighting, water pumping, domestic, military, space and accommodated in grid-connected systems [3].

Mathematical Modeling of Photovoltaic Array

PV modules consist of PV cells connected in series then series- parallel combination of PV module form PV arrays [4]. Series resistance is associated with the losses in the current path due to the metal grid, contacts, and current collecting bus. Shunt resistance due to the loss associated with a small leakage of current through a resistive path in parallel with the built-in device and due to the p-n junction is not ideal. Since shunt resistance is larger than series resistance. The equivalent circuit of the PV cell is shown in Figure 1[5].

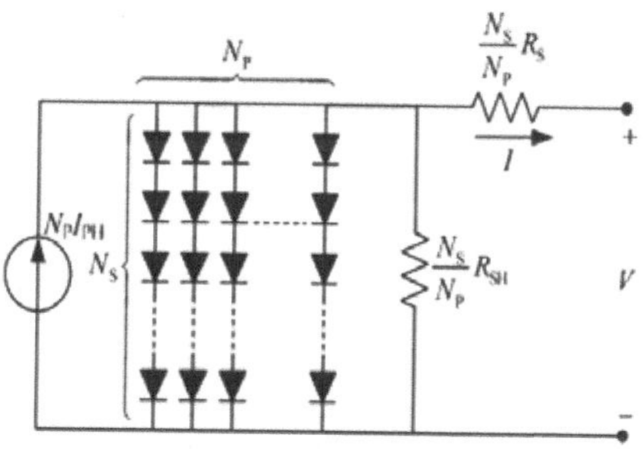

Fig. 1: Equivalent Circuit of Solar Array[2]

Photo-current I_{ph}

$$I_{ph} = [I_{sc} + k_i(T - 298)] \times Ir/1000 \quad \text{... (1)}$$

Reverse saturation current I_{rs},

$$I_{rs} = I_{sc}/\left[\exp\left(V_{oc} \times \frac{q}{N_s} \times k \times n \times T\right) - 1\right] \quad \text{... (2)}$$

Saturation current I_0,

$$I_0 = I_{rs}[T/T_r]^3 \exp[q \times E_{g0}/n \times k(1/T - 1/T_r)] \quad \text{... (3)}$$

PV shunt current I_{sh},

$$I_{sh} = (V \times N_P/N_s + I \times R_s)/R_{sh} \quad \text{... (4)}$$

PV output current,

$$I = I_{ph}N_P - I_0 N_P[\exp(V/N_s + I \times R_s/N_P)q/n \times k \times T - 1] \quad \text{... (5)}$$

Table 1: PV Parameters[4]

Parameters	*Values*
Rated power	100 W
Voltage at maximum power	18 V
Current at maximum power	5.55 A
Open circuit voltage	20.5 V
Short circuit current	6.11 A
Total number of cells in series	36
Total number of cells in parallel	1

Step by Step Procedure for Modeling of Photovoltaic array

The procedure of mathematical modeling has been explained in step by step[4]. A model of PV system includes the temperature independence of the photo-current source, the saturation current of the diode, and a series resistance is considered based on the Shockley diode equation.

Step 1: Input Parameters for Modeling

I_{ph} is photo-current (A); k_i is short-circuit current constant = .002; T is operating temperature (K); Ir is solar irradiation (W/m^2); q is electron charge = 1.602×10^{-19} J/V; n is the ideality factor of the diode = 1.2; K is equal to Boltzmann's constant, = 1.381×10^{-23} J/K; T_r is nominal temperature = 298.15 K; E_{g0} is band gap energy of the semi-conductor = 1.1eV; R_s is series resistance (Ω) = .0001 Ω; R_{sh} is shunt resistance (Ω) = 1000 Ω.

Step 2: Photo-Current i_{ph}

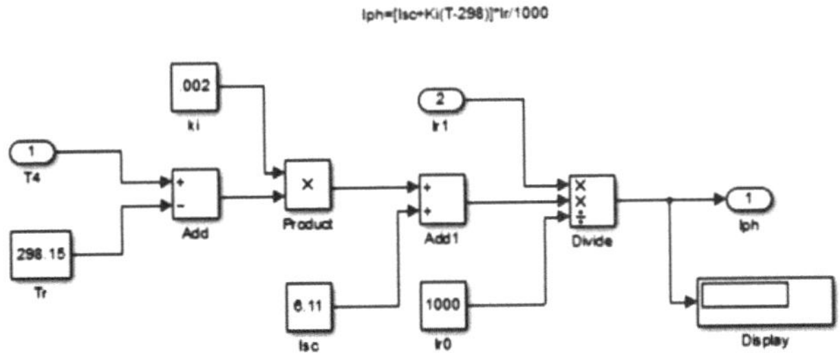

Fig. 2: Photo-Current

Step 3: Reverse Saturation Current i_{rs}

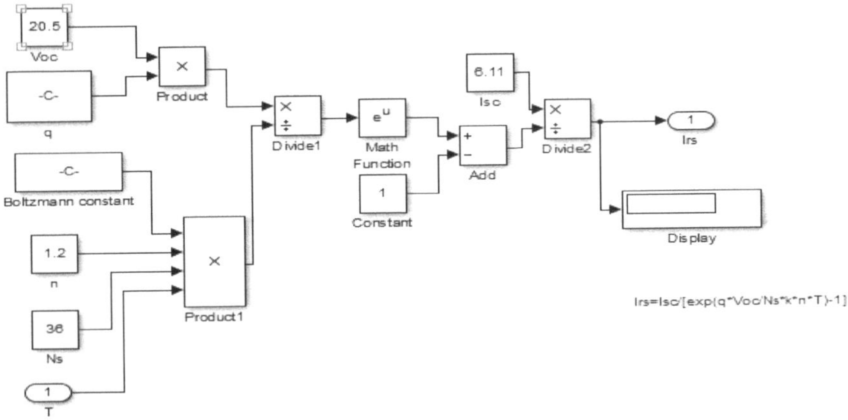

Fig. 3: Reverse Saturation Current

Step 4: Saturation Current i_0

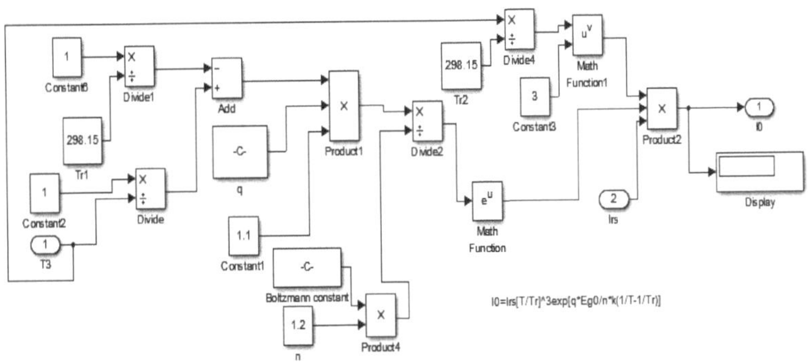

Fig. 4: Saturation Current

Step 5: PV Shunt Current I_{sh}

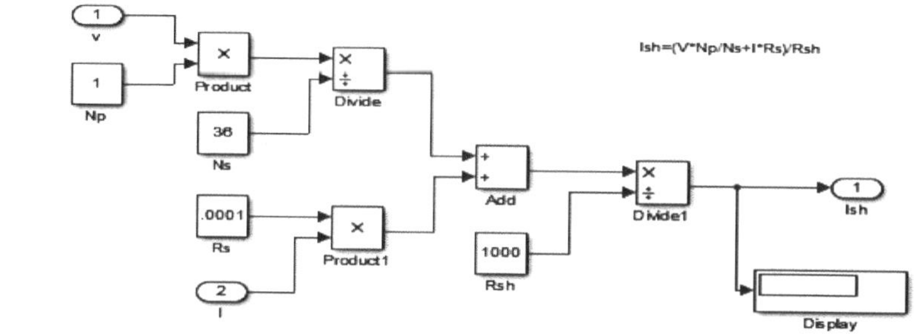

Fig. 5: PV Shunt Current

Step 6: PV Output Current

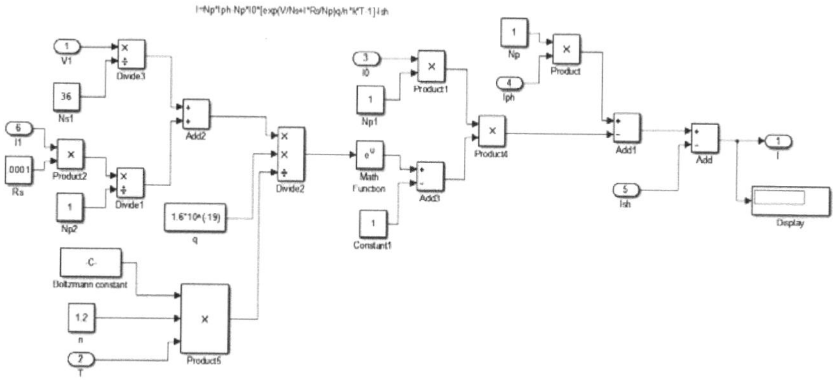

Fig. 6: PV Output Current

Step 7: Interconnection of All Subsystems

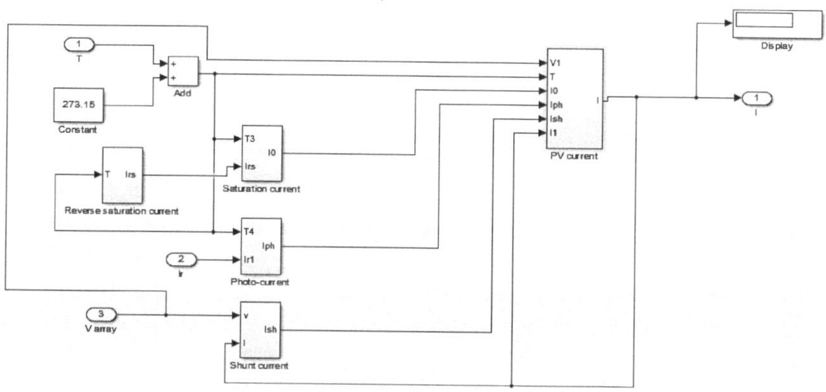

Fig. 7: Interconnection of All Subsystems

Step 8: Solar Subsystem Model

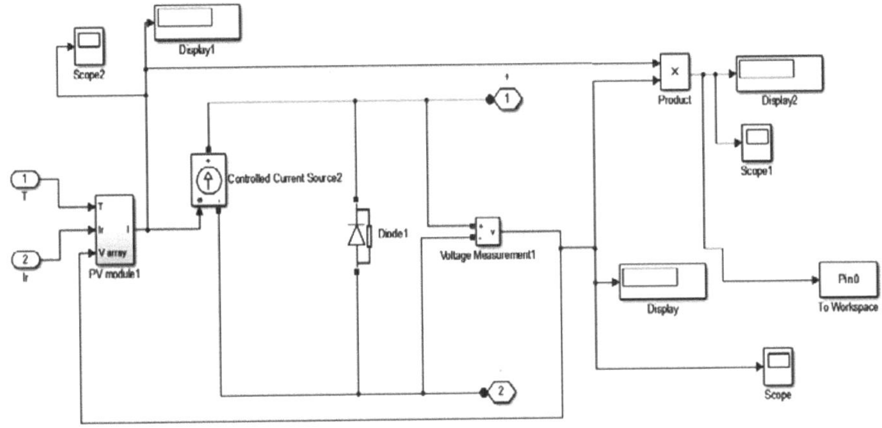

Fig. 8: Solar Subsystem Model

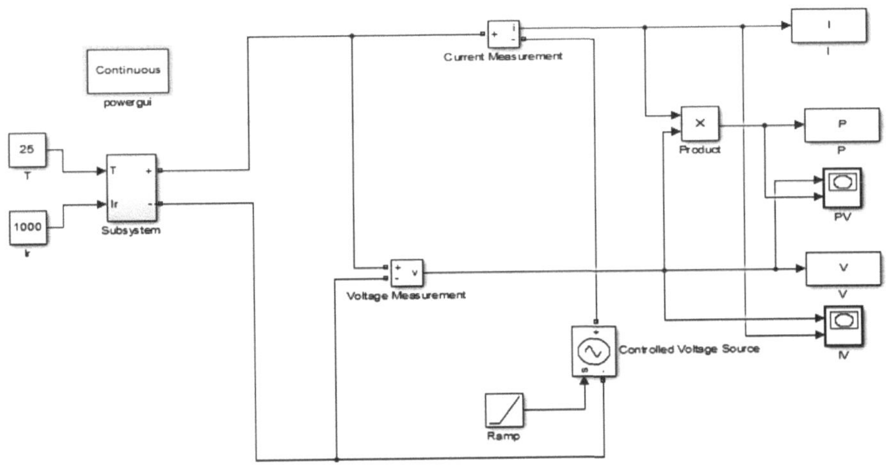

Fig. 9: Solar Photovoltaic System for P-V and I-V Characteristic

Results

The characteristics of PV cell have been explained in two steps. First step is to go to the V-I characteristics of the cell and locate the current corresponding to the voltage at maximum power point. This current is called the current at maximum power point.

Second step is to plot voltage and power graph of the cell. Power is calculated by multiplying voltage across the cell with corresponding current through the cell. From the plot, maximum power point is located and corresponding voltage is noted.

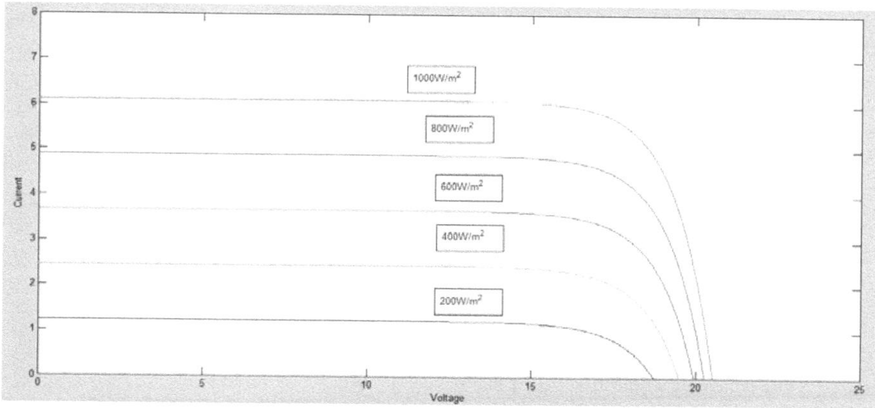

Fig. 10: Current-Voltage Characteristics of Photovoltaic Array

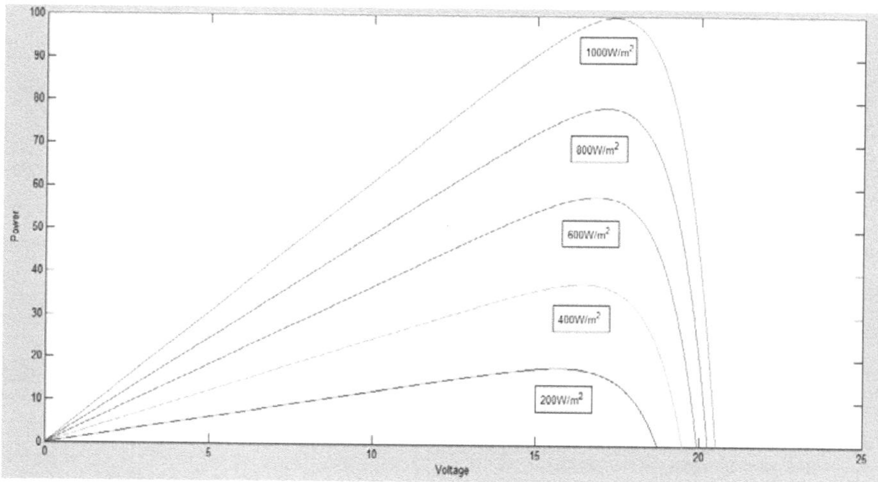

Fig. 11: Power-Voltage Characteristics of Photovoltaic System

Conclusion

A step-by-step procedure for simulating of mathematical modeling of a PV system is carried out in MATLAB/ Simulink. This modeling procedure will help new researcher to understand I–V and P–V operating characteristic of PV array. It is a robust tool to predict the behavior of any solar PV cells, modules and arrays under varying environmental conditions (temperature and irradiation) and physical parameters (series resistance, shunt resistance, ideality factor etc.).

References

[1] Dhar, S., Sridhar, R. and Avasthy, V., "Modeling and Simulation of Photovoltaic Arrays," *Int. J. Eng. Dev. Res.*, Vol. 2, No. 4, pp. 1–5, 2010.

[2] Pandiarajan, N. and Muthu, R., "Mathematical Modeling of Photovoltaic Module with Simulink," *Int. Conf. Electr. Energy Syst.*, Vol. 3, No. 5, pp. 314–319, 2011, doi: 10.1109/ICEES.2011.5725339.

[3] Sheik, M.S. and Devaraj, D., "System with Boost Converter using," *Int. Conf. Circuit, Power Comput. Technol. [ICCPCT]*, Vol. 7, No. 14, pp. 814–821, 2014.

[4] Nguyen, X.H. and Nguyen, M.P., "Mathematical modeling of photovoltaic cell/module/arrays with tags in Matlab/Simulink," *Environ. Syst. Res. a Springer J.*, Vol. 4, No. 24, pp. 2–13, 2015, doi: 10.1186/s40068-015-0047-9.

[5] Noman, A.M., Addoweesh, K.E. and Mashaly, H.M., "A Fuzzy Logic Control Method for MPPT of PV Systems," *IEEE*, Vol. 5, No. 12, pp. 874–880, 2012, doi: 10.1109/IECON.2012.6389174.

Maximum Power Tracking with Perturb and Observation Tecchniques using Single Ended Primary Inductor Converter

Rajat Kumar[1] and Ram Niwash Mahia[2]
Department of Electrical Engineering,
National Institute of Technology Hamirpur, Hamirpur–177005, India
E-mail: [1]20mee011@nith.ac.in; [2]ram@nith.ac.in

ABSTRACT: *Now a days due to increase number of Population, Vehicles on road are also increased and due to which the non-renewable energy sources are decreasing and because of which community are going towards renewable energy source. So this paper proposed a system in which PV model is used to charge battery of electric vehicles using Single Ended Primary Inductor Converter by using this model transportation cost can also be reduced. Due to this irregularities in climatic condition PV model has nonlinear power characteristics. Earlier buck boost and boost converter are used in charge controller, In these converters some drawbacks of boost converter can only step up the voltage and buck converter can only step down voltage this issues output voltage is not constant. In this paper we have used SEPIC converter to provide constant voltage. In this model we have a PV based charging stations are finding enormous capacity and control to charge electric vehicles. The PI controller is used with the SEPIC converter to maintain the duty cycle to be more efficient. In the proposed system MPPT P&O algorithm is written in a MATLAB functional block diagram and implemented in the simulation.*

Keywords: Maximum Power Point Tracking, Perturb & Observation Algorithm, PI Controller, SEPIC Converter.

Introduction

Due to increases in demand of energy use of solar power is also increasing. Solar panel receives sun rays and converts solar energy but the output power of solar panel depends on irradiance and temperature. Irradiance and temperature depends on climate conditions at standard conditions temperature (T) = 25°C, irradiance (Ir) = 1000 w/m^2 it provide rated output. When weather is bad it receives less irradiance and due to which its output power decreases. To overcome this problem maximum power tracking point (MPPT) controller are made [1-2]. The MPPT charge controller allow PV module to operate at maximum power point. To track this maximum power point several algorithm are used like incremental conductance, Perturb & observation and fuzzy logic. In this paper perturb & observation algorithms are used. The MPPT based P&O method is widely used because of running time is less and effective and overall cost is also less [5]. In this paper PV panel is directly fed to dc-dc sepic converter and the advantage of using sepic converter its input and output voltage are same so there no need conversation of voltage [3-4]. Sepic converter is more efficient than other converters. PI controller are used it maintain duty cycle [9-10].

In this paper shows that the maximum power tracking through solar PV panel by using P&O with SEPIC converter [12-13]. The MPPT algorithm is used to maintain the output voltage of PV remains constant [6]. In P&O method is to modify the operating voltage and current of PV panel until we can achieve the maximum power from it, by using modified P & O algorithm the overall proposed system cost is less and efficiency is also improved [8-7].

Now day by day increase in Electric Vehicles are coming on the road, so the charging of EVs is very big challenging for our government so the purpose of this proposed system design to devolved a charging station to charge Electric Vehicles [14]. This purpose of this system designing the overall cost is less as compare to other systems [15]. In Section 2 block diagram representation of proposed system and the basic circuit of PV energy conversion system also discussed. How to increase energy conversion from PV panel and their performance has been discussed in Section 3. The MPPT techniques are discussed in Section 4. The power electronics dc-dc converter and its performance discussed in Section 5.After implementation of proposed system results and conclusion are discussed in section 6.

Proposed System

The overall proposed system shows that Figure 1 it consists of DC source, SEPIC converter, PV panel, MPPT, and PI controller.

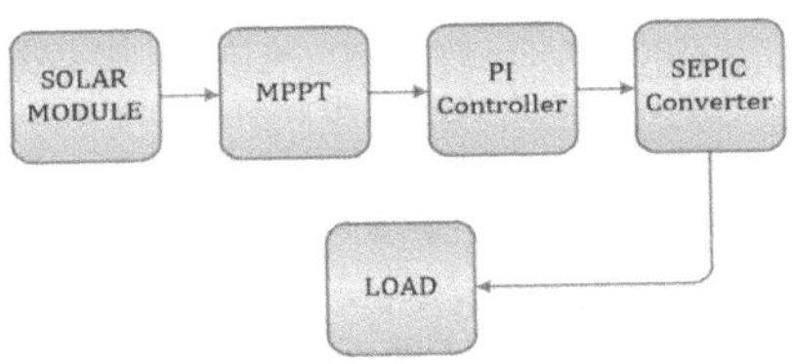

Fig. 1: Block Diagram Representation of Proposed System

In Figure 1 shows that the input is from PV panel and directly fed to SEPIC converter. The input is totally depending upon the atmospherics condition the change in temperature and solar irradiance i.e. the output of Photovoltaic is always fluctuated. The advantage of SEPIC converter is that it works as a buck and boost converter. When output voltage of PV panel is

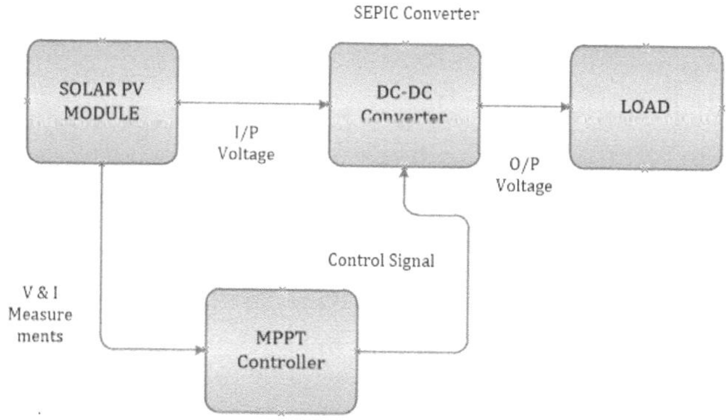

Fig. 2: Basic Circuit of the PV Energy Conversion System

low then the SEPIC converter work as a boost converter and keep maintain constant voltage similarly when PV panel voltage is high at that time the SEPIC converter works as a buck converter and maintain constant voltage and also protect the system from damage. Maximum power point tracking controllers are a fundamental part of photovoltaic systems and algorithms are used to maximize the available power of PV units in a variety of environmental conditions.

The Figure 2 shows that the PV energy conversion in the system. There are three main components solar panel, MPPT controller and dc-dc converter. The solar module is basically consists solar cells [1-2]. The MPPT controller also measures the current and voltage from PV solar panel. For the requirement of voltage at load the system set a Vref voltage for comparing with feedback get desired voltage. The control signal sends a signal to SEPIC converter for comparing voltage and current through MPPT controller. Once output voltage is achevied then this process is stop.

Modeling of PV Cell

The Figure 3 shows that the equivalent circuit of the PV the module. The I_{ph} represents photon current, Id is diode current, and Rs & Rsh are the series and shunt resistances respected, the value of Rs is very small and *R*sh value is very large. The PV module is formed by no of cells connecting in parallel and series [3-4].

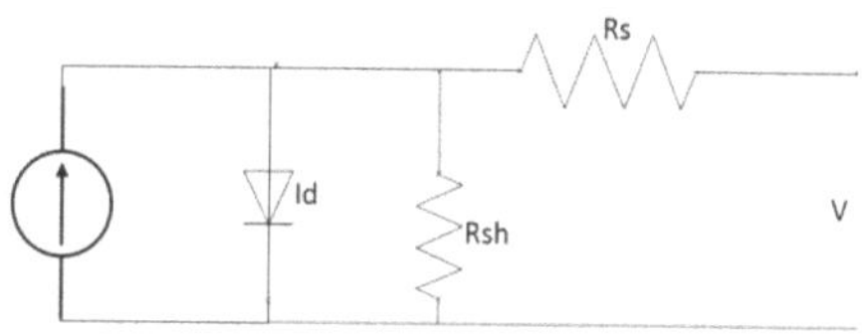

Fig. 3: Equivalent Circuit Configuration of PV Cell

The diode equations can be written as,

$$I_d = I_o \left(\frac{qVd}{e^{akT}} - 1\right) \quad \dots (1)$$

Where,

T = Temperature (k)

q = Charge of electron in columns (1.6×10^{-19})

k = Boltzmann constant (1.38×10^{-23} J/K)

I_o = Reverse saturation current of diode at T

a = Diode Ideality factor

The ideal PV cell can be modelled as,

From equation (1),

$$I = I_{ph} - I_o \left(\frac{qVd}{e^{akT}} - 1\right) \quad \ldots (2)$$

Where,

I_{ph} = Photon current at a given irradiance and given temperature (T), and

V = Voltage across the diode

Now the complete PV model can be defined as,

$$I = I_{ph} - I_d - I_{Rsh} \quad \ldots (3)$$

Put the value of Id from equation (1) and we get the actual current,

$$I = I_{ph} - I_o \left(e^{\frac{q(V+R_S)}{aKT}}\right) - 1\left(\frac{V+IR_s}{R_{sh}}\right) \quad \ldots (4)$$

Designing of PV Panel

The PV panel designing the no of modules required in series & parallel and power. For better understanding assumed that all modules have identical parameters [5–7].

Let,

N_s = Number of Modules used in series

N_p = Number of Modules used in parallel

There are following steps for designing PV Cells.

Step 1: The requirement of current, voltage and power in PV array.

Where,

P_{MA} = Power of PV array.

V_{MA} = Voltage of PV array at maximum power point.

I_{MA} = Current of PV array at maximum power point.

Step 2: To calculate the number of modules to be connected in series and parallel [9].

Where Ns is defined as,

$$N_s = \frac{\text{Total array voltage}}{\text{Voltage of individual module}}$$

Since, PV module is supposed to work under Standard Test Conditions (STC).

Where (STC) is standard test condition Temperature (T) = 25°C and irradiance (I_r) = 1000 w/m^2

$$Pv_{(STC)} = \frac{\text{Array voltage at } V_{MA}}{\text{Module voltage at maximum power point } V_M}$$

Similarly, for parallel string,

$$N_P = \frac{\text{Total array current}}{\text{Current of individual current}}$$

At, STC

$$N_P = \frac{I_{MA}}{I_M} \quad \ldots (5)$$

Step 3: To Calculate total power of the PV array.

Assume,

P_M = Maximum power of a single module.

N_s = Number of module connected in series.

N_P = Number of module connected in parallel.

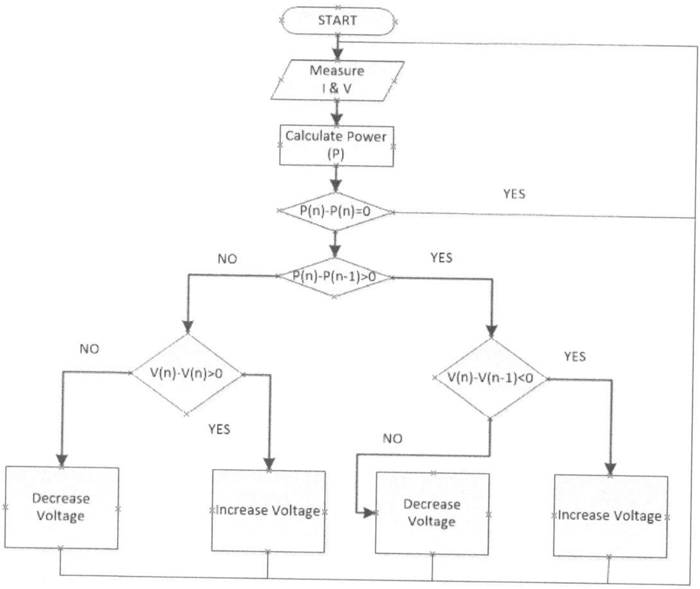

Fig. 4: Flowchart of Perturb & Observe Algorithm

So, total powers can we written as.

$$P_{MA} = N_P \times N_S \times P_M \qquad \dots (6)$$

and

$$\text{Total Power} = V_{MA} \times I_{MA} \qquad \dots (7)$$

MPPT and Perturb & Observe (P&O) Algorithm

The MPPT and P&O algorithm is shown in Figure 4 In this algorithm the output of PV array is totally depended on the temperature and irradiations level of individual PV cells [5]. The P&O method is measures maximum power at different- different point. Let consider three cases for better results regarding maximum power [9].

For achieving Maximum power.

$$\frac{\Delta P_{PV}}{\Delta V_{PV}} = 0 \qquad \dots (8)$$

If the PV panel working at point 1 which is shown in Figure 5 left hand side. At point 1 basically goes down the power would be decreasing and the voltage also decreasing and duty cycle is also increases.

For achieving maximum power at point 1 the voltage is increase and duty cycle is decrease to get maximum power, in this process the SEPIC converter work as a boost converter.

$$\frac{\Delta P}{\Delta V} < 0 \text{ For } V < Vpp \qquad \dots (9)$$

Similarly, in Figure 5 at point 2 goes below the power will decrease and voltage also decrease and duty cycle is increases.

For point 2 goes above the power will increase and voltage is also increase and duty cycle is decreases.

$$\frac{\Delta P}{\Delta V} > 0 \text{ For } V > Vpp \qquad \dots (10)$$

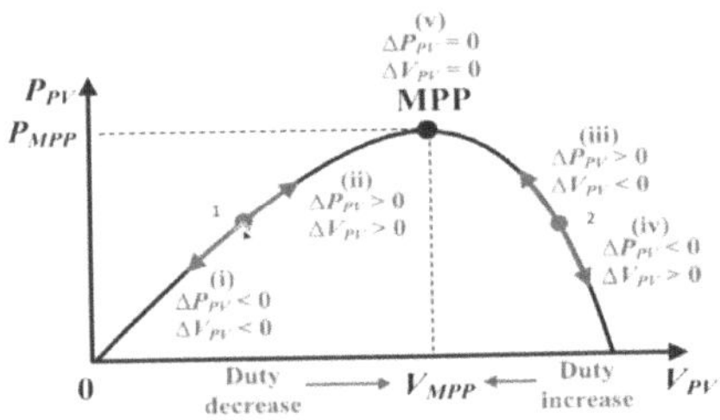

Fig. 5: Perturb & Observe Algorithm PV Curve [17]

For achieving maximum power point is depend on the PV array variation according to the temperature and irradiation. In section V dc-dc sepic converter are used it allows a circuit to have adjustable input voltage with a constant voltage [15].

DC-DC Single Ended Primary Inductor Converter (SEPIC)

The polarity of input voltage and output voltage of SEPIC converter are same which makes it more efficient. The SEPIC converter can be step up and step down the input voltages [9-10].

Working of SEPIC Converter

The working of SEPIC converter there are some following cases are.

Case 1: When switch is on.

In Figure 6 initially the capacitor C1 is fully charged and inductor L1 is charge through input dc voltage and capacitor C1 act as a voltage source and it starts charging the inductor L2.

Case 2: When switch is off:

In Figure 6 inductor L1 is fully charge and reversed it polarity because inductor does not allow sudden change in current and capacitor C1 starts charging through inductor L1 also inductor L2 reverse its polarity and continue to supply current in same direction as when switch is On. Now due to the polarity changed of inductor L2 diode gets forward biased and capacitor C2 starts charging. When C2 gets fully charged irrespective of the switching operation it gets discharged through load [12].

The output voltage of SEPIC converter is.

$$V_O + V_d = \frac{DV_i}{1-d} \quad \dots (11)$$

Designing of Sepic Converter

The duty Ratio (D) defined as,

$$D = \frac{V_O}{V_O + V_s} \quad \dots (12)$$

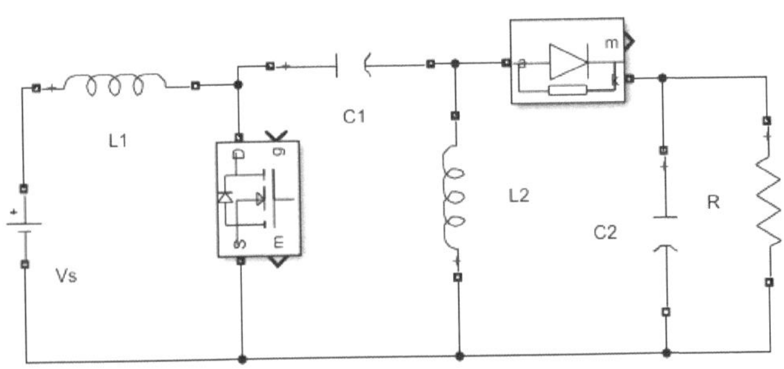

Fig. 6: Equivalent Circuit Configuration of SEPIC Converter

Inductor Current (IL_1) can be written as,

$$IL_1 = \frac{P}{V_S} \quad \ldots (13)$$

Similarly, Inductor current (IL_2) can be written as,

$$IL_2 = \frac{P}{V_0} \quad \ldots (14)$$

$$IL_{1MAX} = IL_1 + \frac{\Delta IL_2}{2} \quad \ldots (15)$$

Minimum Inductor Current (IL_{1Min}) can be written as,

$$IL_{1MIN} = IL_1 - \frac{\Delta IL_1}{2} \quad \ldots (16)$$

Maximum Inductor Current (IL_{2MAX}) can be written as,

$$IL_{2MAX} = IL_2 + \frac{\Delta IL_2}{2} \quad \ldots (17)$$

Minimum Inductor Current (IL_{2MIN}) can be written as,

$$IL_{2MIN} = IL_2 - \frac{\Delta IL_2}{2} \quad \ldots (18)$$

Load Resistance (R) can be written as,

$$R = \frac{V_O^2}{\mathrm{P}} \quad \ldots (19)$$

Capacitance (C_1) can be written as,

$$C_1 = \frac{V_O D}{R_f \times \Delta V_{C1}} \quad \ldots (20)$$

Capacitance (C_2) can be written as,

$$C_2 = \frac{V_O D}{R_f \times \Delta V_{C2}} \quad \ldots (21)$$

Simulation and Results

The simulation of shown in Figure 7 In this proposed system a SEPIC converter is directly connected to PV solar. The SEPIC converter designing parameters are given in Table 1. In this paper we observed that power of PV at atmospheric changing in temperature and irradiance, we also observed that power of PV with and without connecting of MPPT as shown in Figure 8. In this PV system we chose an 1STH-250-WH PV panel module and the maximum power of PV panel is 250.20 W and the module parameter is shown in Table 2.

Table 1: PV Module Paramerters

Parameters	*Value*
Maximum Power	250.2
Series connected module per strings	4
Parallel connected strings	5
Short circuit current Isc (A)	8.66
Shunt Resistance Rsh (ohms)	224.18
Series Resistance Rs (ohms)	0.237
Open circuit voltage Voc (V)	37.3

Table 2: SEPIC Converter Parameters

Parameters	Value
Inductor L1 (H)	0.120
Inductor L2 (H)	0.120
Capacitor C1 (F)	4.28
Capacitor C2 (F)	4.28
Resistance (ohms)	6.4
Output Voltage (V)	178.8

Table 3–4 represents the performance of output power at different-different irradiance and constant temperature. When PV panel is not connected through MPPT controller the performance of output power is less as compare to when PV panel is connected though MPPT controller.

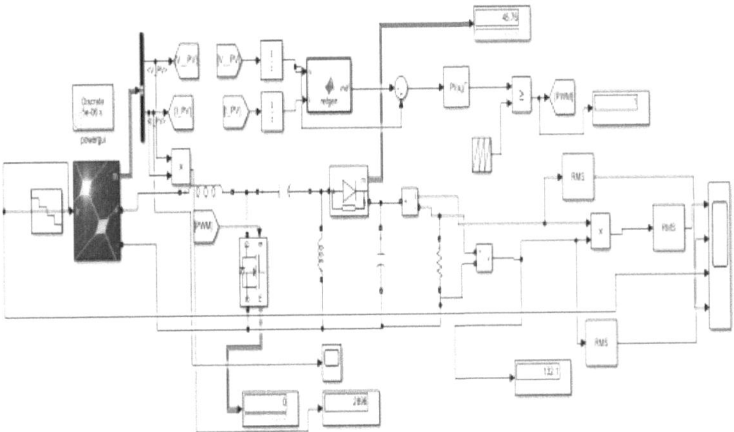

Fig. 7: Simulation Diagram of the Proposed System

Table 3: Performance of Output Power at Without Connecting MPPT Controller at Different Irradiance Condition and Constant Temperature

Parameters	*Irradiance (w/m²)*	*Output Power (Watt)*	*Output Voltage (V)*	*Output Current (A)*
1.	1000	3903	158.1	24.71
2.	900	3770	155.4	24.78
3.	800	3578	151.6	23.68
4.	700	3290	145.2	22.70
5.	600	2731	132.1	20.65

Table 4: Performance of Output Power at with Connecting MPPT Controller at Different Irradiance Condition and Constant Temperature

Parameters	*Irradiance (w/m²)*	*Output Power (Watt)*	*Output Voltage (V)*	*Output Current (A)*
1.	1000	4863	178.8	27.78
2.	900	4116	164.2	25.66
3.	800	3676	155.1	24.33
4.	700	3305	147.2	23.68
5.	600	2821	134.2	21.01

In Figure 8 is shows that the without connecting MPPT controller the PV genearted voltage that is SEPIC input voltage is let to change with irradiance 1000 W/m^2 and constant tempreature, at time = 0.5 sec to be irradiance 1000 W/m^2 and 900, 800, and to 600 at t = 5 sec shows that the output power is less as compare to while MPPT controller conntected with PV panel it is shown in Figure 9. The voltage and current waveforms of SEPIC converter in represents in Figure 9, it is seen regulated traking the given 50 V reference and congicutive to change of atmospheric conditions.

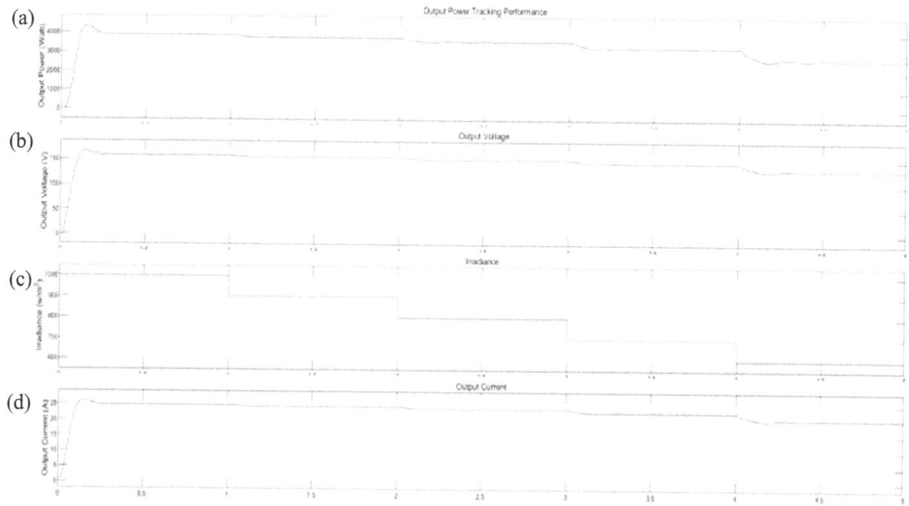

Fig. 8: Waveforms of (a) Output Power of Without Connecting MPPT Controller (b) Output Voltage and Current (c) Irradiance (d) Constant Temperature

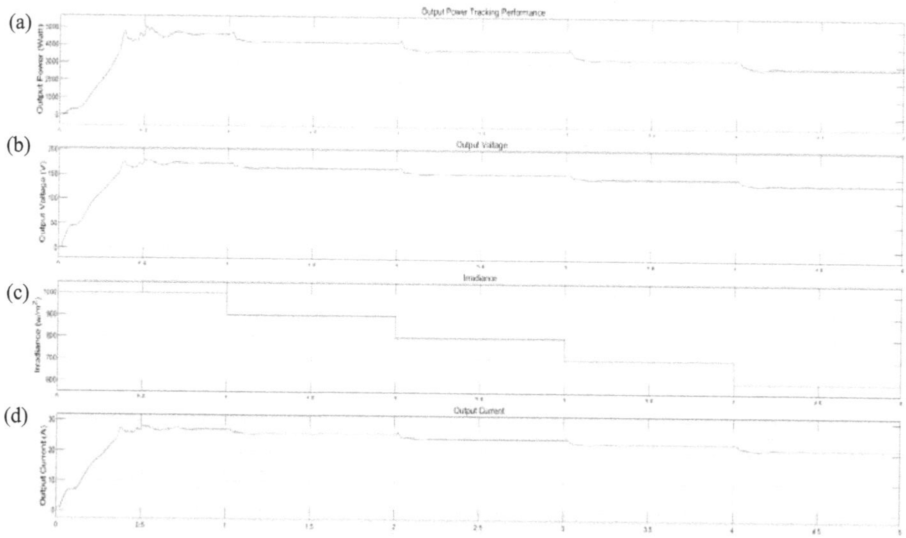

Fig. 9: Waveforms of (a) Output Power of with Connecting MPPT Controller (b) Output Voltage and Current (c) Irradiance (d) Constant Temperature

Conclusion

In this paper presented a MPPT controller and P&O algorithm based SEPIC converter designed for achieving the maximum power from solar panel. The constant voltage of PV systems required to ensure output voltage and output power regulation for any irradiance and temperature conditions. This proposed system is design for achieving maximum output power from PV panel and in this proposed system we observed while connecting PV panel with MPPT controller and without connecting MPPT controller at different-different irradiance and temperature. The purpose of using SEPIC converter it provides large voltage transformation ratio larger then input voltage and also reduce loss across switch. These PV with MPPT controller with modified P&O algorithm with SEPIC converter these type of combination provide better energy conversion, easy implementation abolish the intricacy of control. The efficiency of the overall system is 97%. For further research connect a battery with SEPIC converter for regulated constant voltage for electric vehicles charging station.

References

[1] Al-Rubaye; Mnati, Mohannad Jabbar; Araujo, Victor Gino Morais; Abed, Jameel Kadhim and Bossche, Alex Van den. "Review different types of MPPT techniques for photovoltaic systems." In *International Conference on Sustainable Energy and Environment Sensing (SEES 2018)*, 2018.

[2] Pakkiraiah, B. and Sukumar, G. Durga. "Research survey on various MPPT performance issues to improve the solar PV system efficiency." (2016).

[3] Geethalakshmi, B. and Sreeram, P. "Modeling and analysis of an integrated PV array and SEPIC converter." In *2014 International Conference on Control, Instrumentation, Communication and Computational Technologies (ICCICCT)*, pp. 76–80, IEEE, 2014.

[4] Mishra, Sandeep; Ziar, Hesan; Isabella, Olindo and Zeman, Miro. "Selection map for PV module installation based on shading tolerability and temperature coefficient." *IEEE Journal of Photovoltaics* 9, No. 3 (2019): 872–880.

[5] Kineavy, Fearghal and Duffy, Maeve. "Smart EV charging system for maximising power delivery from renewable sources." In *2016 IEEE Energy Conversion Congress and Exposition (ECCE)*, pp. 1–8. IEEE, 2016.

[6] Singh, Harshit Kumar and Kumar, Nishant. "Solar PV Array Powered ON Board Electric Vehicle Charging with Charging Current Protection Scheme." In *2020 IEEE International Conference on Power Electronics, Drives and Energy Systems (PEDES)*, pp. 1–5. IEEE, 2020.

[7] Shariff, Samir M.; Alam, Mohammad Saad; Ahmad, Furkan; Rafat, Yasser; Asghar, M. Syed Jamil and Khan, Saadullah. "System design and realization of a solar-powered electric vehicle charging station." *IEEE Systems Journal* 14, No. 2 (2019): 2748–2758.

[8] Mazumdar, Poornima; Enjeti, Prasad N. and Balog, Robert S. "Analysis

and design of smart PV modules." *IEEE Journal of Emerging and Selected Topics in Power Electronics* 2, No. 3 (2013): 451-459.

[9] Praba, B. Lakshmi and Seyezhai, R. "Simulation and Hardware Implementation of Interleaved SEPIC Converter with Valley-Fill Circuit for HBLED System." *Trends in Mechanical and Biomedical Design* (2021): 335–353.

[10] Duranay, Zeynep Bala. "Output Voltage Regulated Sepic Converter for Photovoltaic Systems." In *2021 13th International Conference on Electronics, Computers and Artificial Intelligence (ECAI)*, pp. 1–5. IEEE, 2021.

[11] Perianayagi, M.; Prabha, N. Rathina and Gnanavadivel, J. "Design and analysis of single switch dual output DC-DC SEPIC converter for PV applications." In *2019 IEEE International Conference on Innovations in Communication, Computing and Instrumentation (ICCI)*, pp. 112-116. IEEE, 2019.

[12] Karuppiah, M., Dineshkumar, P. and Karthikumar, K. "Design a Electric Vehicle Charger Based Sepic Topology with PI Controller." In *2020 IEEE International Conference on Advances and Developments in Electrical and Electronics Engineering (ICADEE)*, pp. 1–5. IEEE, 2020.

[13] Sunddararaj, Suvetha Poyyamani; Rangarajan, Shriram S.; Subramaniam, Umashankar; Collins, E. Randolp and Senjyu, Tomonobu. "Performance of P/PI/PID Based controller in DC-DC Converter for PV applications and Smart Grid Technology." In *2021 7th International Conference on Electrical Energy Systems (ICEES)*, pp. 171–176, IEEE, 2021.

[14] Uniyal, Jagriti; Joshi, MonaLisa and Juneja, Pradeep K. "Set point tracking capability and comparitive response of PI-PD controller with PI, PD and PID controller." In *2016 International Conference on Advances in Computing, Communication, & Automation (ICACCA) (Spring)*, pp. 1–6, IEEE, 2016.

[15] Lee, Hyeon-Seok and Yun, Jae-Jung. "Advanced MPPT algorithm for distributed photovoltaic systems." *Energies* 12, No. 18 (2019): 3576.

Performance Evaluation of Grid Connected Wind Farm with Variation in Wind Speed

Santosh Kumari Meena[1], M.K. Bhaskar[2] and Khamma Kanwar[3]

[1,3]M.B.M. Engineering College, Jai Narain Vyas University, Jodhpur–342001, India

[2]Department of Electrical Engineering, M.B.M. Engineering College, Jai Narain Vyas University, Jodhpur–342001, India

E-mail: [1]sanumeena11245@gmail.com; [2]MKB932@gmail.com; [3]kkjoya.jnvu@gmail.com

ABSTRACT: *The worldwide consumption of electrical energy is increasing and there is gradually increased of demand on the power generation. Thus in accumulation to conventional power generation sources a various number of renewable energy sources are being connected into the power system. The wind turbine power generation unit is more economical and competitive of all the existing environmentally clean and safe renewable energy sources. However, wind electrical generation systems provide intermittent energy depending upon the wind regimes in the region. Therefore, any transient disturbance can result in changes in the voltage, current and power at various points of the integrated system. This paper shows simulation results for variation in wind speed obtained through transient analysis using generic wind turbine generation model. In the wind farm a doubly fed induction generator (DFIG) model is used. For showing the simulation of DFIG, MATLAB/SIMULINK software is used.*

Keywords: Doubly Fed Induction Generator (DFIG), Point of Common Coupling (PCC), Wind Energy Conversion System (WECS), Pulsed Width Modulated (PWM), GSC (Grid side converter).

Introduction

The worldwide consumption of electrical energy is increasing and there is gradually increased of demand on the power generation. Thus in accumulation to conventional power generation sources a various number of renewable energy sources are being connected into the power system. Wind is most desirable source surrounded by all other renewable energy because it is non-polluting, available in abundance and reasonable cost for both small and large order system. The most satisfy resource this whole need of wind in a natural source energy source. The wind turbine power generation unit is more economical and competitive of all the existing environmentally clean and safe renewable energy sources in world. In wind turbine generation technology variable speed double fed induction and constant-speed squirrel-cage induction generators are used. However, wind electrical generation systems provide intermittent energy depending upon the wind regimes in the region. Therefore, any transient disturbance can result in changes in the voltage, current and power at various points of the integrated system. A study of such effects is useful to the power system energy planner.

Wind Energy Conversion System

Wind energy conversion system (WECS) is applied to change wind energy into useful mechanical energy. The terms "wind power" or "wind energy" relate the operation by which the wind is used for develop electric energy or mechanical energy. In rural and remote areas, wind turbine system is most generally installed. These areas generally have weak grids and which generally have under voltage conditions and voltage unbalances. The torque generated by the electrical machine (induction generator) is fluting in nature when the stator phase voltage supplied by the grid is unbalanced in nature. The developed torque is pulsations in nature with double the grid frequency and resulted in acoustic noise at low levels and at high levels and it can damage the blade assembly, gearbox or rotor shaft. The induction generator will take the unbalanced current when the grid is unbalanced. These unbalanced current causes over current problems as well increases the grid voltage unbalance. The most important components of a typical WECS are wind turbine, inter-connection apparatus, turbine generator and control systems. The wind-turbine is used as the prime-mover to supply mechanical power to an electrical generator (DFIG). The WECS have basically consists of two considerable components:

1. Turbine system and associated control box (including the gear box).
2. Electrical system consist the rotor converter system and the associated control.

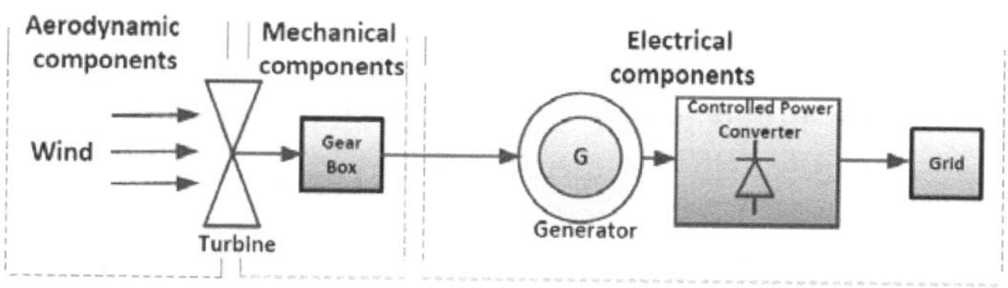

Fig. 1: Block Diagram of the Components of WECS Connected to the Grid

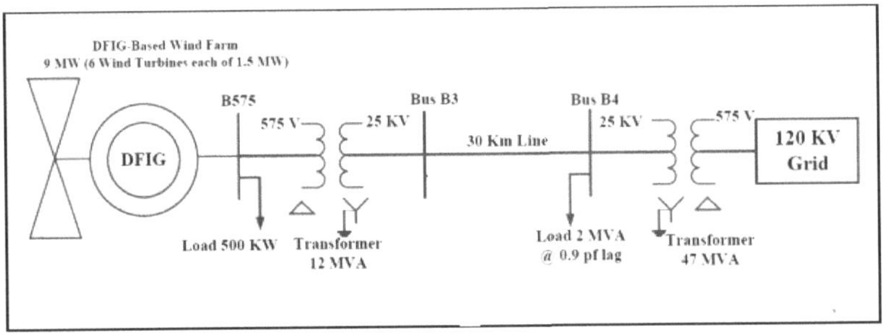

Fig. 2: Single Line Diagram of the Test System

Power Grid Model

The single line diagram of the test system used for the *MATLAB simulation* is shown in Figure 3. *MATLAB/SIMULINK* is used for the analysis. The grid model [American system] consists of a 120 kV, 60 Hz grid supply point, feeding the supply to 25 kV distribution system through a step down transformer of 120–25 kV, 47 MVA, another 25 kV–575 V, 12 MVA step-down transformer connected between grid and bus B4 to again step down the voltage at the level 575 volt.

There are two loads in the system; one load of 2 MVA, 0.9 p.f. (lag) at 30 km from transmission line, and a static load of 500 kW at 575 V bus. The 25 kV, 30 km long line is represented as nominal-Π line. A DFIG-based wind farm consists of six DFIG-based wind turbines connected at 575 V bus, each with 1.5 MW (for a total of 9 MW) which have a protection system monitoring voltage, current, speed and the DC link voltage. In this model, the wind speed increases slowly from 8m/sec and reaches the final constant value of 14 m/sec in 16 sec. All the tests here are studied after the system reaches steady state (i.e. after 16 seconds). The GSC (Grid side converter) in DFIG maintains the DC link voltage to almost constant value at 1,200 V during normal operating conditions .

SimPower System toolbox used for the simulation of wind farm of 9 MW wind farm. Six DFIG-based wind turbines are connected together and represented by a single equivalent wind turbine mode of 9 MW. Each individual DFIG wind turbine system has a capacity of 1.5 MW, which results a total of 9 MW DFIG-based Wind farm. The simulation analysis is discussed using *MATLAB/SIMULINK* platform for various transient disturbance cases.

Induction Generator

Wind turbines mainly consist wound rotor type double fed induction generator and two IGBT based PWM converters. These two converters are connected with DFIG in which one is rotor side and other is grid side. In which any one act as converter for converting AC to DC and other act as inverter for convert DC to AC. In this type of generator, the stator winding is directly connected to grid supply while the rotor is supplied by two converter system of AC/DC/AC at variable frequency. Here the DFIG used have nominal power of 1.67 MVA, 575 Volt (line to line) at frequency of 60 HZ. For grounding purpose at the secondary side (Δ) of grid, a grounding Transformer is applied with capacity of 47 MVA and leakage impedance (Xo = 4).

In Table 1 all the parameters of the generator are represented.

Table 1: Three Phase DFIG Parameters

Parameters	*Symbol*	*Value*
Resistance of stator winding	R_1	0.00474 p.u.
leakage inductance of stator winding	L_1	0.125 p.u
Rotor Resistance ref to Stator	R_2	0.00434 p.u.
Rotor inductance ref to Stator	L_2	0.179 p.u.
Magnetizing Inductance	Lm	6.5 p.u.

Three Phase Transformer

Two transformers with Y_g/Y_n and Y_g/Δ (D1) configuration of three-phase (2-winding) used with a nominal power of 4 MVA and 47 MVA Respectively. The windings parameters for primary and secondary windings are shown in following Table. For providing grounding point a grounding transformer of 47 MVA and X_o = 4.7 is connected to the secondary side (Δ) of grid. All the transformer parameters shown in Table 2.

Table 2: Transformer Parameters

Parameters	*3-Phase 4MVA Transformer (Yg/Yn)*	*47MVA Grid Side Transformer Yg/Δ(D1)*
	Primary Secondary Winding Winding	*Primary Secondary Winding Winding*
Voltage (phase to phase) rms (KV)	33 0.575	132 33
Resistance (R) pu	0.025/30 0.025/30	0.08/30 0.08/30
Inductance (L) pu	0.025 0.025	0.08 0.08

Transmission line

The 25 kV, 30 km long line is represented as nominal-Π line. A short transmission line with length of 1 Km is connected with individual wind generators of 1.5 MW with same parameters specified earlier. Table 3 shows the parameters of the line.

Table 3: Transmission Line Parameters

Parameters	*Positive Sequence*	*Negative Sequence*
Resistance (perKm)	0.1154	0.414
Inductance (mH/Km)	1.052	3.324
Capacitance (μF/Km)	11.34	5.011

Results and Discussion

System Response with Change in Wind Speed

DFIG-based wind farm was initially run under normal operation to calculate the steady state output conditions of the system. Wind speed given as input to a step block that shows the sudden variation in wind speed. At starting the wind speed fixed at 8 m/s and at a time of t = 5s the wind speed is suddenly increased to 14 m/s through a step signal. The control mode is fixed at voltage regulation mode with the help of control block of the DFIG-linked wind farm. In this mode reference voltage is set 1 p.u. on the basis of generator rating which are 6 × 1.5 MVA and V = 575 volt at the point of common coupling (PCC) at bus B-575. At time t = 5s, the generated active power begins increase gradually with the turbine speed and Approximately 18 sec and also reach to its maximum power generation capacity of 9 MW. Within same time the speed of turbine is rise from 0.8 to 1.2 p.u. of the synchronous speed. At PCC (Bus B575 in this model) voltage should be maintain at 1 p.u. so that it is necessary that reactive power should be under controlled. The equivalent DC link voltage should be approximately 1200 volt and increase pitch angle from 0 to 0.78 deg to restrict the mechanical power. At rated power, the wind turbine absorbs 0.682 Mvar reactive power (Q = –0.68 Mvar) to maintain the voltage 1 p.u. equivalent to reference voltage

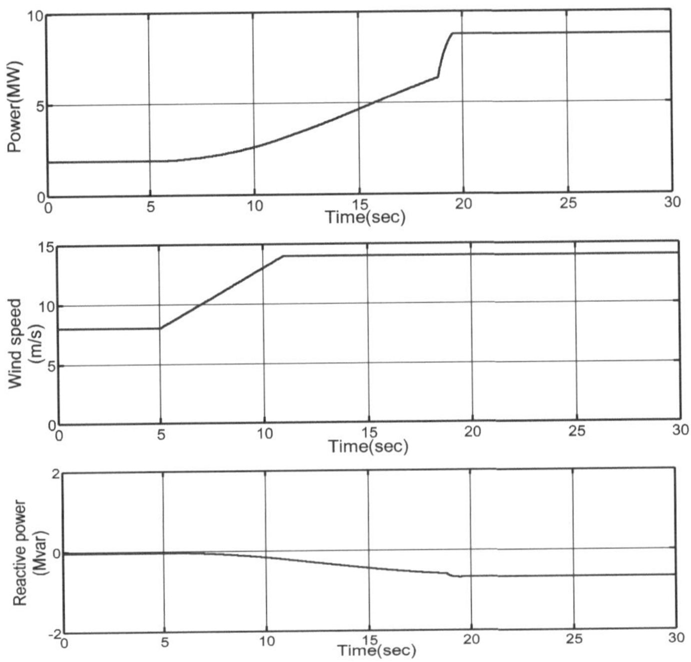

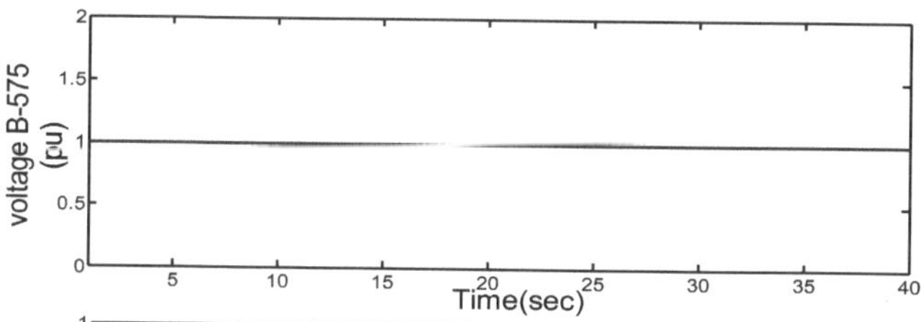

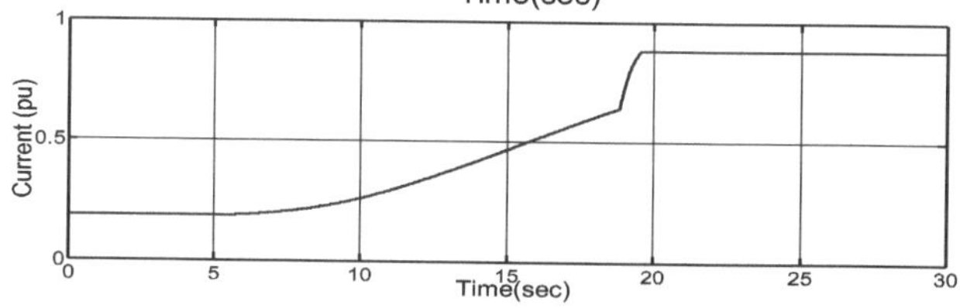

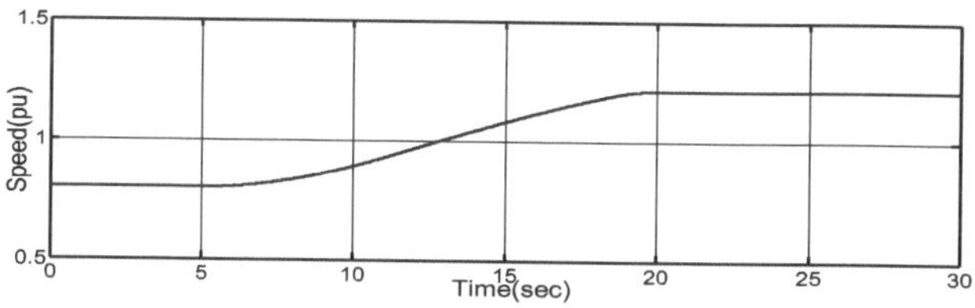

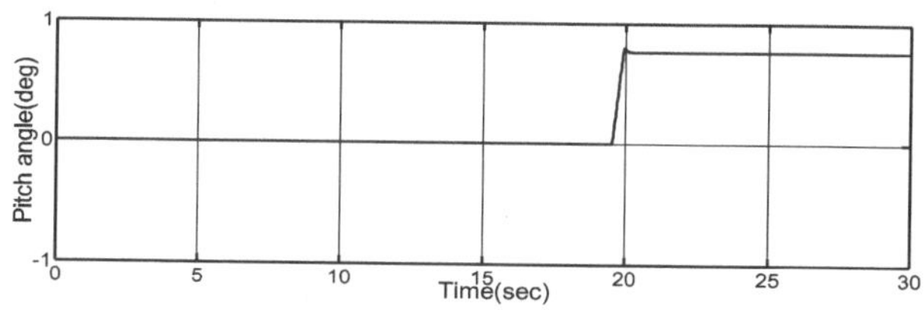

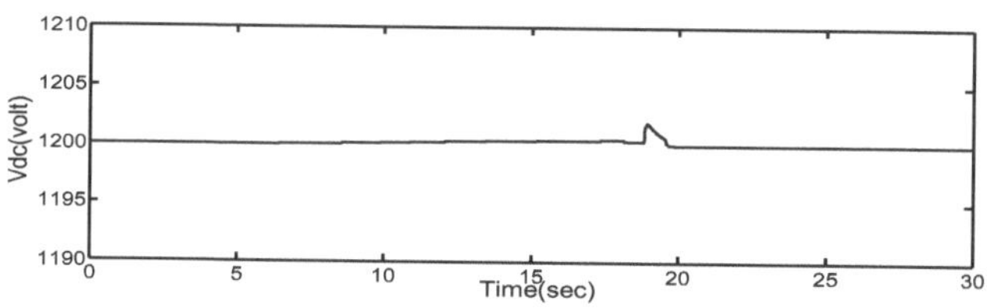

Conclusion

On the basis of the study it can be concluded that the dynamic behavior of variable speed DFIG coupled with disturbances in wind speed in the power system. The real output power of turbine varies with the variation in the base wind speed. Fluctuations of the wind speed will produce a prominent influence on the stability of DFIG. Change in grid voltage is related to change in reactive power demand as at higher grid voltages.

References

[1] Masaud, Tarek; Lee, Keun; Sen, P.K. and Malmedal, Keith, "Study of the Implementation of STATCOM on

DFIG-Based Wind Farm Connected to a Power System", *IEEE Conference, PEDES2010,* New Delhi, India, pp. 1–6, Dec. 20–23, 2010.

[2] Qiao, Wei; Venayagamoorthy, Ganesh Kumar and Harley, Ronald G., "Real-time implementation of a STATCOM on a wind farm equipped with doubly fed induction generators," *IEEE Trans. Industry App.*, Vol. 45, No. 1, pp. 98–107, Feb. 2009.

[3] Engelhardt, Stephan; Erlich, Istvan; Feltes, Christian; Kretschmann, Jorg and Shewarega, Fekadu, "Reactive Power Capability of Wind Turbines Based on Doubly Fed. Induction Generators," *IEEE Transactions on Energy Conversion*, Vol. 26, No. 1, pp. 364–372, March 2011.

[4] Okedu, K.E., Muyeen, S.M., Takahashi, Rion and Tamura, Junji, "Participation of FACTS in stabilizing DFIG with crowbar during grid fault based on grid codes," *IEEE Conference and Exhibition (GCC),* Dubai, UAE, pp. 365–368, February, 19–22, 2011.

[5] Fazli, Mehrdad; Shafighi, Ali Reza; Fazli, Ali and Shayanfar, Heidar Ali, "Effects of STATCOM on Wind Turbines Equipped with DFIGs During Grid Faults," *World Non-Grid-Connected Wind Power and Energy Conference (WNWEC)*, Nanjing, China, 5–7 Nov. 2010.

[6] Jayam, A.P. and Chowdhury, B.H., "Improving the Dynamic Performance of Wind Farms with STATCOM," *IEEE Power System Conference and Exposition, Seattle*, USA, March, 15–18, 2009.

[7] Qiao, Wei and Harley, R.G. "Improved Control of DFIG Wind Turbines for Operation with Unbalanced Network Voltages", *Industry Applications Society Annual Meeting*, 2008, IAS '08. IEEE 5–9 Oct. 2008, pp. 1–7.

[8] Herbert, G.M. Joselin; Iniyan, S., Sreevalsa, E. and Rajapandian, S., "A review of wind energy technologies," *Renewable and Sustainable Energy Reviews*, Vol. 11, pp. 1117–1145, 2007.

[9] Muljadi, Eduard; Butterfield, C.P.; Parsons, Brian and Ellis, Abraham, "Effect of Variable Speed Wind Turbine Generator on Stability of a Weak Grid," *IEEE Transactions on Energy Conversion*, Vol. 22, No. 1, pp. 29–36, March 2007.

[10] Mohan, Ned and Brekken, Ted K.A. "Control of a Doubly Fed Induction Wind Generator Under Unbalanced Grid Voltage Conditions", *IEEE Transaction Energy Conversion*, Vol. 22. 1, March 2007, pp. 129–135.

[11] Muljadi, E., Butterfield, C., Parsons, B. and Ellis, A., "Effect of variable speed wind turbine generator on stability of a weak grid," *IEEE Transactions on Energy Conversion*, Vol. 22, No. 1, pp. 29–36, 2007.

[12] Zhou, Y. and Bauer, P., "Control of DFIG under Unsymmetrical Voltage dips", *Power Electronics Specialists Conference, 2*007. IEEE 17–21, June 2007, Page(s): 933–938.

[13] Jang, Jeong-Ik; Kim, Young-Sin and Lee, Dong-Choon, "Active and Reactive Power Control of DFIG for Wind Energy Conversion under Unbalanced Grid Voltage," *CES/ IEEE 5th International Power Electronics and Motion Control Conference (IPEMC),* Shanghai, China, Vol. 3, pp. 1–5, Aug. 2006.

Predicting IMDb Rating of Movies by Machine Learning Technique

Shailendra Singh Jakhar[1] and Chandershekhar Singh[2]
Department of Electrical Engineering, Jodhpur Institute of Engineering and Technology, JIET Universe, NH-62, Pali Road, Mogra, Jodhpur–342802, India
E-mail: [1]shailu200112@gmail.com; [2]chandershekhar333@gmail.com

ABSTRACT: *The film industry is a world of dreams. It is not only industry or a career of entertainment, but rather it is now a center of global business, with movies earning billions of dollars. The world is now excited about a movie's success and popularity. A huge amount of data is available online about these movies' success and ratings. We used the Hollywood movie list from Wikipedia and their rating from the IMDb movie rating website to create our data set. Then machine learning classification algorithms are applied to the data set. The goal is to build a regression model to predict the IMDb rating of movies using features that are only available prior to their release*

Keywords: Machine Learning; IMDb Rating; Beautiful Soup; Linear Regression; Prediction.

Introduction

Now-a-days, movies are not a source of recreation; instead, they are one of the primary sources of global commerce and marketing. The film industry is a profession that most people automatically write off as impossible to get inside. It seems like one of those careers where "It's not what you know, it's whom you know," and you do not know anyone. Creative professions are indeed tough to get in. Unlike more business-minded fields, like accountants or lawyers (who know what they must do to succeed), the film industry has no guarantees and no clear checklist to victory. Film professionals are entirely on their own when finding a path to success, and it can be very intimidating, especially since no two career paths are alike. There are many studies to predict the movie success rate as people are too much excited about films. Few studies include the movie's features (director, screenplay, actor, actress, genre) to predict the success rate. So, in this paper, the focus is to predict the success rate based on the movie's features—for example, Director, Screenplay, Actors/Actress, Country, and Genre.

Again, IMDb is a popular platform that is growing day by day. Generally, people show much interest if the movies get the highest vote [10, 12]. Therefore, the ultimate motive is to find a model that can efficiently predict the IMDb Rating using only available features prior to its release.

Related Work

Several works predict the movie success rate before a movie is released. Many researchers have used various machine learning techniques to predict the success rate. Data about movie attributes rather than social media data is analyzed, and logistic regression gives an 84% accuracy level [3]. Pramod, Abhisht, and Geetha show that Fuzzy logic gives high accuracy for categorizing predictions [4]. Machine learning approaches are applied to the synthetic dataset to build an efficient prediction structure using IMDB [3]. Various machine learning algorithms are being used to predict movie success rates [5]. Mathematical models are implemented to determine movie success depending on various movie attributes [6]. Another study shows the market shares of domestic and international movies in Russia, where the international movies are ahead from 2002 to 2014 [2]. This paper also distinguished three factors behind the success of a movie. They are budget brands like famous actors/actresses, directors, and viewers' reviews. Lastly, the model concludes that the sanction and budget have a comparatively high effect on the success of movies. An exciting work about analyzing movie-goers' tastes and behaviour was done recently. Here, individuals' taste is first analyzed to make a model then aggregated to predict box office results [7]. The Indian film industry is vast, and its impact and influence are enormous. So, data mining techniques are used to predict whether a Bollywood movie will be a blockbuster one or not [1]. Many studies are done based on social media data analysis [8]. The data may be a review, comments, or reactions to an event.

Problem Domain

This paper focuses on finding a model to predict a movie's IMDb rating using the movie's information like the studio, director, screenplay, actor, actress, genre, and the country as input features. In the data set, the title of the movie and the year of the movie release are also stated. The model is built using machine learning techniques. It is trained and tested by the data set prepared from Wikipedia and IMDb movie rating website [1]. Hence, the model built here is learned and tested by a supervised learning technique. The data set created here contains the class attribute "rating," which can be a flop, below average, average, and hit, as measured by the rating from the IMDb website. A list of Hollywood movies released in 2018 is fetched from Wikipedia. Then the ratings of each of those movies are collected one by one from the IMDb website. Therefore, the data set is an actual set and contains data from two different sources. Then several popular machine learning classification algorithms are applied to the data set. Nevertheless, as the data set is imbalanced, techniques are used to balance it before applying classification algorithms. Resampling without replacement is done to satisfy accuracy measures with the data set. In its entirety, this project explored a few critical skills required of a data scientist:

- Web scraping (requests, HTML, Beautiful Soup)
- EDA (pandas, numpy)
- Linear regression (scikit-learn)
- Data visualization (seaborn, matplotlib).

Methodology

The working method for this work involves a few steps. The methodology is shown in Figure 1. The steps are described below.

- Data Extraction (Acquisition)
- Data Processing (Cleaning)

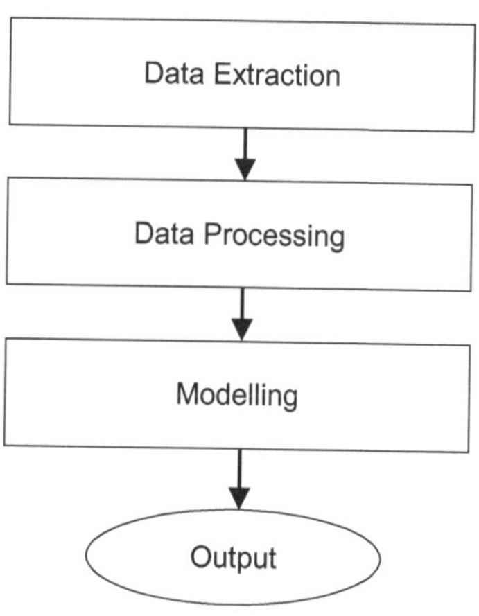

Fig. 1: Working Methodology

- Applying ML Techniques
- Comparing the result of different algorithms

Data Acquisition

The information was gathered from IMDb using requests and Beautiful Soup. *Requests* are the module required to take the webpage and turn it into an object in python. Beautiful Soup takes that object, which is the HTML information behind the webpage, and makes searching and accessing specific information within the HTML text easy. An advanced search feature was used to access titles between 2000 and 2020. The results spanned thousands of pages, and each page held the titles and links to 100 movies. Upon further inspection, it was noticed that the URL contained the phrase: 'start=1'. Increasing this start number by 100 would flip through each page. With a helper function, requests and Beautiful Soup were used to pull the links for each page and return a list of those links.

A function was created to extract data to utilize that list of movie hyperlinks from each page. This function took in a link and returned a dictionary containing the following information: title, IMDb rating, the number of IMDb rates, MPAA rating, genres, directors, writers, top three stars, initial country of the release, original language of the release, release date, budget, opening weekend USA, gross USA, cumulative worldwide gross, production companies, and runtime [9].

Data Cleaning

Some data had to be cleaned as part of the Exploratory Data Analysis (EDA). It consisted of turning any numerical value from a string into an integer. Runtime had to be converted into minutes, all monetary values needed commas and dollar signs removed, and the release date had to be converted into Date Time. Additionally, categories that contained lists needed to be converted from strings into actual python lists (genres, directors, stars, production companies) [9]. The retrieval function did most of this cleaning, but some other cleaning was necessary after putting the data into a Data Frame.

More processing was done with over 2,000 movies in a Data Frame to get an available Data Frame for modelling. The movies with a budget under $1000, no. of votes less than 1500 were dropped. Regarding that last requirement, movies with a few rates reported more extreme ratings (movies leaning towards a perfect ten or a big goose egg).

Pair Plots: Pair plots are a great visualization tool for exploring relationships within the data and informing where to start for an MVP. It seems like much information, but when you format your Data Frame with the first or last column being the target, it is much easier to interpret all of this information. The plots in the first column show relationships between the independent variables and the target for this pair plot.

Modelling

It is important to note that another requirement for this project was linear regression, so the models we experimented with were linear regressions and ridge regressions.

Firstly, a simple linear regression was done with the runtime exclusive feature and IMDb rating as the target. It resulted in an R^2 value of 0.2687.

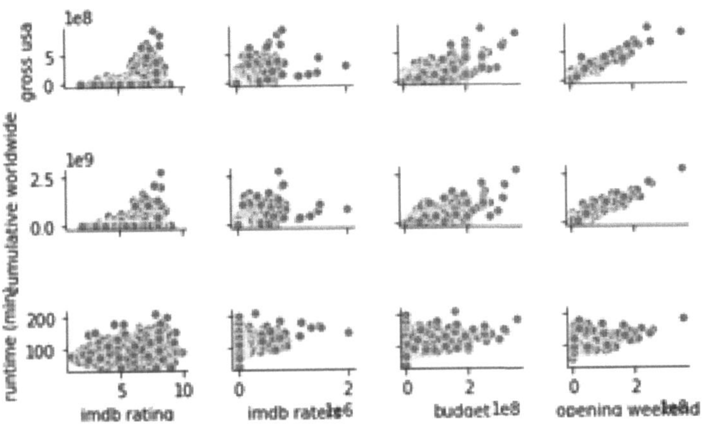

Fig. 2: Data Visualisation using Pair Plots

For MPAA rating and genre, dummy variables were created to add to the Data Frame and got an R^2 of 0.3997. As for directors, writers, stars, and production companies. A list of the most frequently occurring players in each category was created as dummy variables for the top contenders. If a director only appeared once in data, then that director's weight (or coefficient) would directly result from that specific film's rating, so having players with multiple rows of data would give the model more information to create a better-informed coefficient. The release date was used to make a 'release month' feature. Having loaded the features into a model, a resulting R^2 of 0.4751 seemed promising, but the next step was to rigorously test the model with cross-validation.

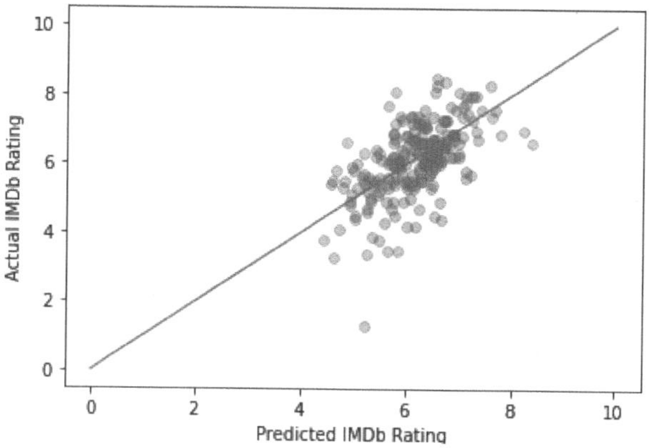

Fig. 3: Ridge Regression Model *vs.* the Target

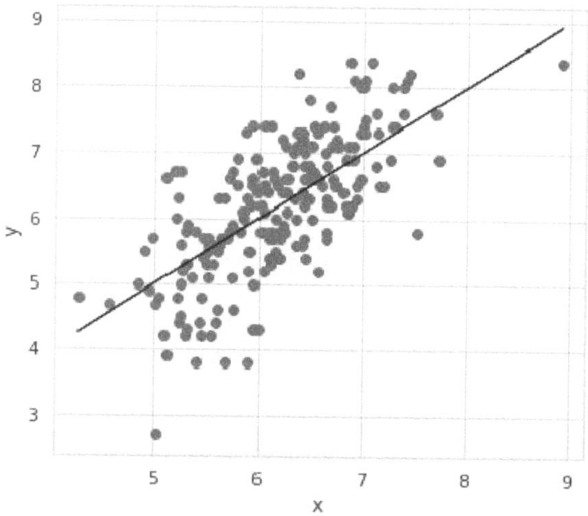

Fig. 4: Regression Fit

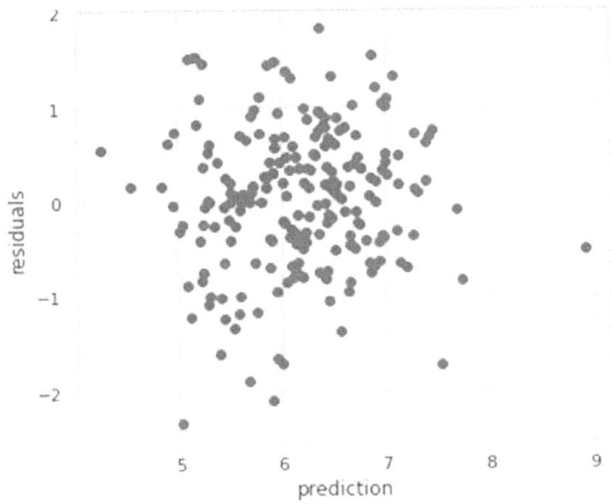

Fig. 5: Residual Plot

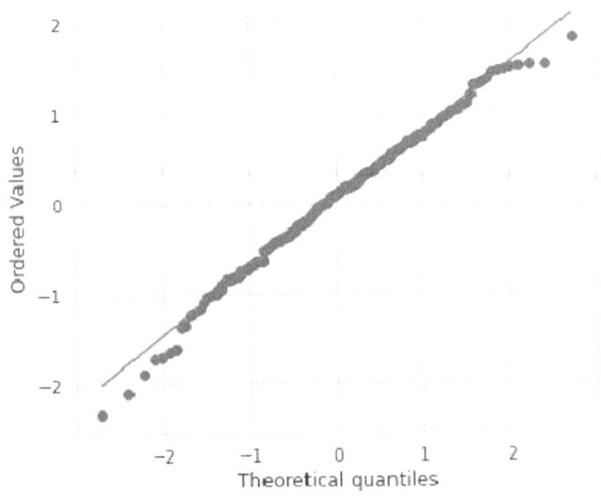

Fig. 6: Normal Q-Q Plot

Results

The final model resulted in an R^2 of 0.432 and a mean absolute error of 0.64. It is a reasonably low R^2, but this article describes why an R^2 below 0.5 for predicting human behaviour is expected. Additionally, the plot to the left of predicted ratings vs actual ratings provided more confidence in the model, as the model has a linear relationship. Also, the movies with the highest residuals had either a low number of ratings or were movies like Cats, Fifty Shades of Grey, and The Emoji Movie. These movies have good stats, but the public did not receive them well, which is a complex metric to incorporate into this model.

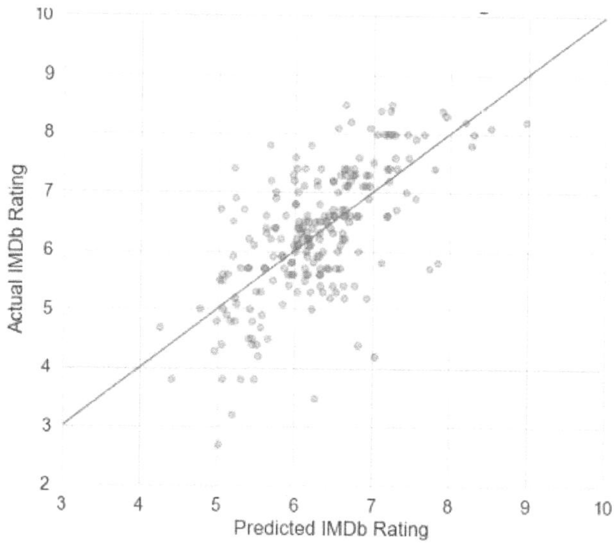

Fig. 7: Final Model Prediction *vs.* Target

It is also essential to look at the coefficients associated with each feature. That is the beauty of ridge regression: using the coefficients to determine the weight of a specific feature.

In the end, the model predicted an IMDb rating with an R^2 of 0.432, significantly better than just predicting with the mean and an MSE of 0.64, which means the prediction was liable to be wrong by 0.64 points in either direction.

Conclusion

As the business market of film industries is becoming massive day by day, competition here is also growing complex. Therefore, predicting a movie's rating is growing complex also [1]. Our model is developed on a real-world data set, and it is collected from two platforms, Wikipedia and IMDb. The model can also be used to predict some other ratings like Rotten Tomato or Metacritic. Other than films, TV shows, music shows can be predicted by our model using the features of our model. [2] Some researchers describe some features having more influence on movie success and others having less or no influence. According to those researches [2], budget has a slight positive influence, but the actor/actress does not influence the Russian film industry. Thus, our work can be done by a weighted feature classify-cation to reflect these influences.

Bollywood or another movie dataset can make the model more efficient with the Hollywood movie dataset. The database can be enriched by including the movies released in recent years. Along with 2018, movies from 2017 or 2016 can be included.

References

[1] Bristi, Warda Ruheen; Zaman, Zakia and Sultana, Nishat, "Predicting IMDb Rating of Movies by Machine Learning Techniques", 2019.

[2] Gaenssle, S., Budzinski, O. and Astakhova, D., "Conquering the box office: Factors influencing success of international movies in russia," 2018.

[3] Latif, M.H. and Afzal, H., "Prediction of movies popularity using machine learning techniques," *International Journal of Computer Science and Network Security (IJCSNS),* Vol. 16, No. 8, p. 127, 2016.

[4] Pramod, S., Joshi, A. and Mary, A., "Prediction of movie success for real world movie dataset," *Int. J. of Advance Res., Ideas and Innovations in Technol*, Vol. 3, No. 3, 2017.

[5] Parimi, R. and Caragea, D., "Pre-release box-office success prediction for motion pictures," in *International Workshop on Machine Learning and Data Mining in Pattern Recognition*. Springer, 2013, pp. 571–585.

[6] Ahmad, J., Duraisamy, P., Yousef, A. and Buckles, B., "Movie success prediction using data mining," in 2017 *8th International Conference on Computing, Communication and Networking Technologies (ICCCNT), IEEE*, 2017, pp. 1–4.

[7] Ruhrländer, R.P., Boissier, M. and Uflacker, M., "Improving box office result predictions for movies using consumer-centric models," in *Proceedings of the 24th ACM SIGKDD International Conference on Knowledge Discovery & Data Mining.* ACM, 2018, pp. 655–664.

[8] Çizmeci, Beyza and Ögüdücü, Şule Gündüz, "Predicting imdb ratings of pre-release movies with factorization machines using social media," in *2018 3rd International Conference on Computer Science and Engineering (UBMK),* IEEE, 2018.

[9] Cowell, Joe, "Predicting IMDb Movie Ratings Using Supervised Machine Learning" at *https://towardsdatascience.com/predicting-imdb-movie-ratings-using-supervised-machine-learning-f3b126ab2ddb.*

[10] Saraee, M., White, S., Eccleston, J. *et al.*, "A data mining approach to analysis and prediction of movie ratings," *Transactions of the Wessex Institute.*

[11] Han, J. and Kamber, M., "Data Mining Concepts and Techniques", Morgan Kaufmann Publishers: San Francisco, 2001.

[12] Neurosoft S.A., Neurosoft Envisioner, www.neurosoft.gr/products/envi.asp, 1999.

[13] Thearling, K., "Data Mining and Analytic Technologies", www.thearling.com, 2004.

[14] Hamilton *et al.,* "Knowledge Discovery in Databases", www2.cs.uregina.ca/~hamilton/courses/831/, 2002.

[15] Attar Software Ltd., Active Data Mining Solutions, www.attar.com/tutor/deploy.htm, 2004.

[16] Labovitz, M.L., "What Is Data Mining and What Are Its Uses?", www.darwinmag.com/read/100103/mining.html, 2003.

[17] Nautilus Systems Inc, "The Data Mining Process", www.nautilussystems.com/process.html, 1996.

Fractional Order Controller based Performance Enhancement of Electric Vehicle System

M. Manjula[1] and C. Srisailam[2]

[1]Department of Electrical Engineering, OU College of Engineering, Hyderabad, Telangana, India

[2]Department of Electrical and Electronics Engineering, Chaitanya Bharathi Institute of Technology (A), Hyderabad, Telangana, India

E-mail: [1] manjulagooga@gmail.com; [2]srisailamc_eee@cbit.ac.in

ABSTRACT: *Most of the electrical and mechanical systems whose transfer function can be found by the manipulating the mathematical equations. The typical values are substituted in this transfer function and a suitable controller is applied like conventional PID controller, any other controller. This gives satisfactory results. They are accepted by industry. However better results always possible. Hence fractional order controller is applied which are given superior results than the conventional controller.*

In this paper fractional order controller is applied for electrical vehicle and performance has been obtained and results are discussed.

Keywords: Fractional Order Controller; PID Controller; Electrical Vehicle Performance Indices; Motor Drive.

Introduction

Fractional order calculus has a growing industry due to its wide application resources [1]. In recent years there are numerous systems make use of fractional order calculus to get superior results than any integer order calculus. For any differential solution an integer order solution exists but fractional order solution does not [2-3]. Therefore, to know the effect of a fractional order controller on a transfer function one must first convert an integer order transfer function to a fractional order transfer function and then fairly accurate this fractional order transfer function to an integer order transfer function using Oustapp or any other related methods.

At present there is cumbersome task of realizing Fractional order controller as a model [4, 5]. But simulation of the same can be done to get better results. Currently industry is a booming one where researches are trying hard to acquire desired results. So it would not be long to be able to make a model of the same. As of now the study of observability, controllability & stability of the fractional order systems are the latest trending topics dealt by researchers.

Controlling industrial plants requires satisfaction of wide range of specifications. So, wide ranges of techniques are needed. Mostly for industrial applications, integer order controllers are used for controlling purpose. Now day's fractional order (FOPID) controller is used for industrial application to improve the system control performances [4-7]. The most common form of a fractional order PID controller is the controller. FOPID controller provides extra degree of freedom for not only the need of design controller gains but also design orders of integral and derivative [8, 9]. The orders of integral and derivative are not necessarily integer, but any real numbers. As shown in Figure 1, The FOPID controller generalizes the conventional integer order PID controller and expands it from point to plane. This expansion could provide much more flexibility in PID control design. The transfer function of such a controller has the following form.

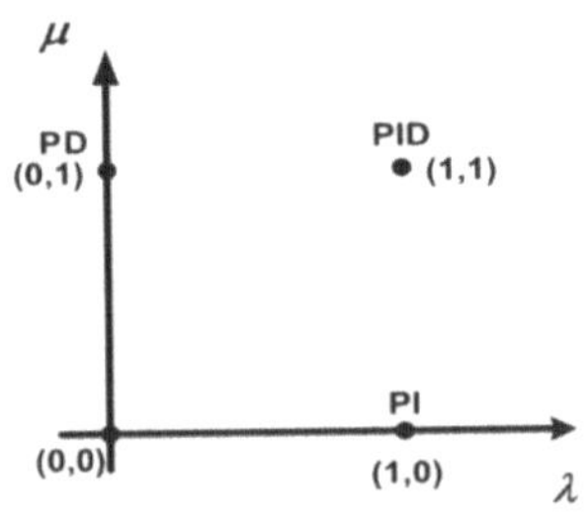

Fig. 1: General form of a Fractional Order PID Controller

Methodology

Fractional-order calculus is defined as:

$$ {}_aD_t^\alpha = \begin{cases} \dfrac{d^a}{dt^\alpha}, \text{Re } \alpha \rangle 0 \\ 1, \text{Re } \alpha = 0 \\ \int_a^t (d\tau)^{-\alpha}, \text{Re } \alpha \langle 0 \end{cases} \quad \dots (1) $$

Where, '*a*' & '*t*' are the limits of the operation and 'α' denotes the order of fractional calculus. Here, α can be a real or a complex value [4]. The generalized fractional-order differential equations (FODEs) is given as,

$$ a_n D^{\alpha_n} y(t) + a_{n-1} D^{\alpha_n - 1} y(t) + \dots\dots a_0 D^{\alpha_0} y(t) $$
$$ b_n D^{\beta_n} u(t) + b_{n-1} D^{\beta_n - 1} u(t) + \dots\dots b_0 D^{\alpha_0} u(t) \quad \dots (2) $$

Where, $(a_i, b_j) \in R^2$ and $(\alpha_i, \beta_j) \in R^2$. The fractional-order transfer function that represents the process model having an input delay term is,

$$ G(S) = \frac{b_n S^{\beta_n} + b_{n-1} S^{\beta_n - 1} + \dots\dots b_0 S^{\alpha_0}}{a_n S^{\alpha_n} + a_{n-1} S^{\alpha_n - 1} + \dots\dots a_0 D^{\alpha_0}} e^{-Ls} \quad \dots (3) $$

Where it is usual to take $\beta_0 = \alpha_0 = 0$ so that the static gain of the system is given by $K = \frac{b_0}{a_0}$, and $L \in R$.

Generalized Fraction Order Control

The theorized form of a rational-order PID controller is in the form,

$$G(S) = K_P + K_i S^{-\lambda} + K_d S^{\mu} \quad \dots (4)$$

Where, λ and μ are orders of integration and differentiation, respectively and the operators of integral and derivative are rationalized in nature. The systematic way of changing the values of λ and μ according to the necessity i.e. PI (λ = 1 and μ = 0), PD (λ = 0 and μ = 1), PID (λ = 1 and μ = 1), the various forms of PID controllers can be obtained to meet requirements.

Feedback Control System

The main purpose of designing the control system is so as to meet some criteria so that the output can be set to a fixed value which is called as reference value. Even though there are some unknown disturbances, reference value should be maintained. The first one is said to be tracking, the second one is said to be disturbance rejection and if the condition are met then the control system design can be a robust servomechanism. Automatically reduce errors. Improve stability Increase robustness against external disturbances

Time Domain Analysis

Starting from random initialized parameters progressively minimizes different integral performance indices iteratively while finding optimal set of parameters for the FOPID and PID controller. The algorithm terminates if the value of the objective function does not change appreciably over some successive iterations.

For designing controllers based on time domain, controllers aim at minimization of different integral performance indices namely:

1. Integral square Error (ISE) = $\int_0^{\tau} e^{2}(t)dt$

2. Integral absolute Error (IAE) = $\int_0^{\tau} |e(t)|dt$

3. Integral time - square Error (ITSE) = $\int_0^{\tau} te^{2}(t)dt$

4. Integral time - absolute Error (ITAE) = $\int_0^{\tau} t|e(t)|dt$

Description of EV

Electrical vehicle is a machine, which takes voltage as input and gives mechanical speed as output, the basic energy flow starts from battery pack to power electronic devices and then electric traction motor will be powered by them, the rotor of motor is mechanically coupled to the shaft of the wheels, which in turns develops the motion of vehicle, through on-board charger, battery pack will gets charged and powers the vehicle.

The force-balance equation makes use of Newton's laws of motion by a technique called mathematical modeling and the governed equation as follows.

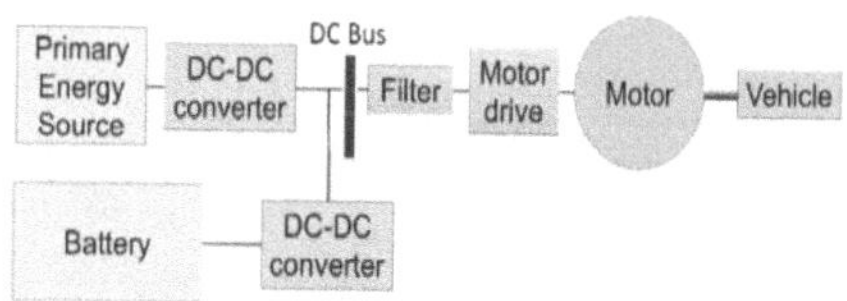

Fig. 2: Block Diagram of Electrical Vehicle

$$F = fmg\cos\alpha + mg\sin\alpha + 0.5\rho C_w A(v + v_{hw})^2 + k_m ma \quad \dots (5)$$

Table 1: Parameters Description and Typical Values

Parameter	*Description*	*Typical Values*
m	Mass of the car in kg	1590 kg
F	Coefficient of rolling resistance	0.011
C_W	Coefficient of aerodynamic drag	0.3
k_m	Coefficient to compensate for apparent increase in vehicle mass to rotating masses	1.2
P	Air density, in kg/m^3	1.2 kg/m^3
A	Largest cross-section of the car, in m^2	2.0 m^2
G	gravitational acceleration, in m/s^2	9.8 m/s^2
A	Inclination of path on which car rides	
V	Velocity of the car	

By using F-V or F-I analogy convert the mechanical in to electrical parameters and obtain the integer order transfer function (IOTF) of the system by substituting the above typical values in it and it as follows.

$$\text{IOTF} = \frac{0.913242}{1.39S^2 + 1.215S + 0.913242} \quad \ldots (6)$$

PID Tuning

The parameters obtained for the PID controller namely k_p, k_i, k_d by Ziegler Nicholas method or Ant colony optimization technique, but in simulation there is a tool called PID tuner which directly gives the parameters when the plant transfer function entered as an input and obtained parameters are as follows.

Table 2: Results Obtained for PID Parameter of System under Consideration

k_p	k_i	k_d
0.6885	0.5454	0.2173

As per the system concern, two gains are available. One from plant and another from controller so by cascading them open loop transfer function build up, then calculate closed loop transfer function of system by using appropriate MATLAB instructions. Obtained results of time domain specifications are as follows.

Table 3: Results Obtained for Time Domain Specifications of System under Consideration

Time Domain Specification	*Value*
Rise time	2.4209s
Settling time	9.3037s
Settling min	0.9050s
Settling max	1.0328s
overshoot	3.2826%
undershoot	0
peak	1.0382
Peak time	4.2995s

Optimization Technique

Generally, any optimization technique is used to obtain the design parameters as per the requirements. Here the use of four different kinds of optimization techniques, the parameters are optimized in simulation process with an extension called FOMCON, which is exceptionally used for simulating fractional order systems, this extension is facilitated with inbuilt optimization algorithms.

Nelder-Mead Method

This is a simplest method devised by Nelder and mead in 1965 for finding a local minimum of a function of several variables. For two variables a simplex is a triangle and the method is a pattern search that compares function values at the three vertices of a triangle. The worst vertex, where functional value is largest is rejected and replaced with new vertex. A new triangle is formed and search is continued. The process generates a sequence of triangles for which functional values at vertices get smaller. The sizes of triangles get reduced and the coordinates of minimum point are obtained.

Interior-Point Method

Interior point methods or barrier methods are a certain class of algorithms to solve linear and non-linear convex optimization problems. Violation of inequality constraints are prevented by augmenting the objective function with a barrier term that causes the optimal unconstrained value to be in the feasible space.

Sequential Quadratic Programming (SQP) Method

Sequential quadratic programming is an iterative method for constrained non-linear optimization. SQP methods solve a sequence of optimization sub problems, each of which optimizes a quadratic model of the objective subject to a linearization of the constraints. If the problem is unconstrained, then the method reduces to Newton's method for finding a point where the gradient of the objective vanishes. If the problem has only equality constraints, then the method is equivalent to applying Newton's method to the first order optimality conditions or karush-Kuhn-tucker conditions of the problem.

Table 4: Results are Obtained for the System with FOPID Controller by Using N-M Optimization Technique

Time Domain Specifications	*ISE*	*IAE*	*ITSE*	*ITAE*
Rise time	0.8774s	0.7546s	0.5938s	0.9769s
Peak time	0.9914s	1.3742s	1.4263s	1.7175s
Over shoot	1.0002	1.8735	32.5291	1.2216
Peak	0.9914	1.0002	1.3179	0.9925
Settling min	0.8043s	0.8145s	0.7728s	0.7932s
Settling max	0.9914s	1.0002s	1.3179s	0.9925s
Settling time	101.01s	99.378s	14.361s	108.02s

Table 5: Results are Obtained for the System with FOPID Controller by Using Interior Point Optimization Technique

Time Domain Specifications	*ISE*	*IAE*	*ITSE*	*ITAE*
Rise time	0.3214s	1.6873s	0.6821s	0.0515s
Peak time	0.6718s	4.2706s	2.0849s	0.1152s
Over shoot	7.8731	14.5707	9.8074	7.2554
Peak	1.0766	1.1451	1.0977	1.0717
Settling min	0.9011s	0.968s	0.9242s	0.9027s
Settling max	1.0766s	1.1451s	1.0977s	1.0717s
Settling time	7.4643s	14.0932s	7.0749s	0.2077s

Table 6: Results are Obtained for the System with FOPID Controller by Using SQP Method

Time Domain Specifications	*ISE*	*IAE*	*ITSE*	*ITAE*
Rise time	0.6139s	0.7928s	0.5216s	0.8923s
Peak time	1.297s	1.8706s	1.0283s	1.070s
Over shoot	5.868	22.614	5.809	7.145
Peak	1.051	1.225	1.057	1.070
Settling min	0.908s	0.938s	0.909s	0.906s
Settling max	1.051s	1.225s	1.057s	1.070s
Settling time	18.719s	5.9917s	4.3118s	2.498s

Active Set Method

The active set is particularly important in optimization theory, as it determines which constraints will influence the final result of optimization. For example, in solving the linear programming problem, the active set gives the hyper plane that intersects at the solution point. In quadratic programming, as the solution is not necessarily on one of the edges of the bounding polygon, an estimation of the active set gives us a subset of inequalities to watch while searching the solution, which reduces the complexity of the search.

Analysis from Optimization Techniques

From the above Table 8 none of the specified optimization techniques produces the entire time domain specifications. Only few parameters are improved in every method. So best of the available methods is NELDER-MEAD method only because in this method maximum two parameters are improved.

Table 7: Results Obtained for the System with FOPID Controller by Using Active Set Optimization Technique

Time Domain Specifications	*ISE*	*IAE*	*ITSE*	*ITAE*
Rise time	0.13s	0.832s	1.628s	0.095s
Peak time	0.288s	1.608s	11.018s	0.2168s
Over shoot	9.003	4.577	10.152	9.3201
Peak	1.087	1.045	1.0993	1.093
Settling min	0.899s	0.919s	0.898s	0.9013s
Settling max	1.087s	1.045s	1.0993s	1.093s
Settling time	0.5334s	0.9196s	22.7821s	NaN

Table 8: ISE with Various OT and Time Domain Specifications

ISE	*Rise Time*	*Over Shoot*	*Peak*	*Settling Time*
N-M	0.8774s	1.0002	0.9914	101.01s
I-P	0.3214s	7.8731	1.0766	7.4643s
SQP	0.6139s	5.868	1.051	18.719s
Active	0.13s	9.003	1.087	0.5334s

Table 9: IAE with Various OT and Time Domain Specifications

IAE	*Rise Time*	*Over Shoot*	*Peak*	*Settling Time*
N-M	0.7546s	1.8735	1.0002	99.378s
I-P	1.6873s	14.5707	1.1451	14.0932s
SQP	0.7928s	22.614	1.225	5.9917s
Active	0.832s	4.577	1.045	0.9196s

From the above Table 9 none of the specified optimization techniques produces the entire time domain specifications. Only few parameters are improved in every method. So best of the available methods is NELDER-MEAD method only because, in this method maximum three parameters are improved. Rise time, peak over shoot and peak time superior in this method compared to all other methods.

Table 10: ISTE with Various OT and Time Domain Specifications

ISTE	*Rise Time*	*Over shoot*	*Peak*	*Settling Time*
N-M	0.5938s	32.5291	1.3179	14.361s
I-P	0.6821s	9.8074	1.0977	7.0749s
SQP	0.5216s	5.809	1.057	4.3118s
Active	0.5216s	10.152	1.0993	22.7821s

Table 11: ITAE with Various OT and Time Domain Specifications

ITAE	*Rise Time*	*Over Shoot*	*Peak*	*Settling Time*
N-M	0.9769s	1.2216	0.9925	108.02s
I-P	0.0515s	7.2554	1.0717	0.2077s
SQP	0.8923s	7.145	1.07	2.498s
Active	0.095s	9.3201	1.093	NaN

From the above Table 10 none of the specified optimization techniques produces the entire time domain specifications. Only few parameters are improved in every method. So best of the available methods is SQP method only because, in this method maximum three parameters are improved. Rise time, peak over shoot and peak time superior in this method compared to all other methods.

From the above Table 11 none of the specified optimization techniques produces the entire time domain specifications. Only few parameters are improved in every method. So best of the available methods is NELDER-MEAD method only because, in this method maximum three parameters are improved. Rise time, peak over shoot and peak time superior in this method compared to all other methods.

Controllers aim at minimization of different integral performance indices. From the above discussion, the NELDER-MEAD method is superior, because in this method most of the errors parameters are improved.

Oustaloup's Approximation Algorithm

The Oustaloup's approximation algorithm is well known method used for integer order approximation of fractional systems where a frequency band of interest is considered (10^{-3} to 10^{3}). Suppose that the frequency range to be fit by an integer order filters to fractional order derivative are given by [ω_b, ω_h], the terms s/ω_u can be substituted with

$$C_0 \frac{1+s/\omega_b}{1+s/\omega_h}$$

Where $\sqrt{\omega_b \omega_h} = \omega_u$ and $C_0 = \frac{\omega_b}{\omega_u} = \frac{\omega_u}{\omega_h}$

The Oustaloup's approximation model to a fractional order differentiator can be written as,

$$G_1(S) = (C_0)^{\alpha} \prod_{k=-N}^{K=N} \frac{1 + s/\omega^*_k}{1 + s/\omega_k}$$

Where the parameters

$$\omega^*_k = \omega_b \left(\frac{\omega_b}{\omega_u} \right)^{\frac{k+N+0.5+0.5\alpha}{2N+1}}$$

and $\omega_k = \omega_b \left(\frac{\omega_b}{\omega_u} \right)^{\frac{k+N+0.5-0.5\alpha}{2N+1}}$ are respectively the zeros and poles of rank k. The total number of zeros or poles is given as $(2N+1)$.

Observations

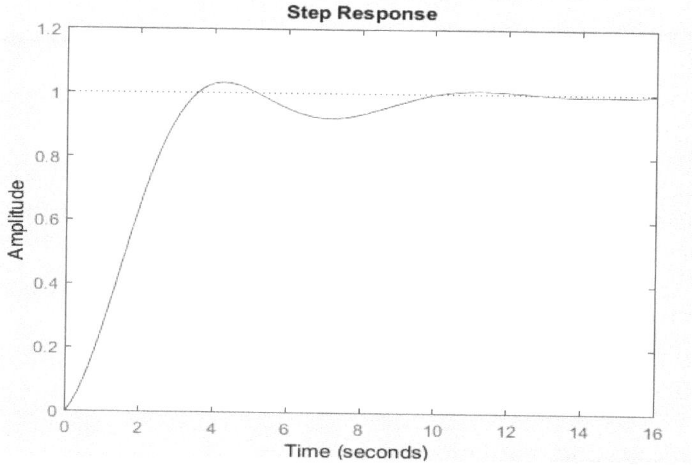

Fig. 3: Step Response of Under Damped Electrical Vehicle System (EVS)

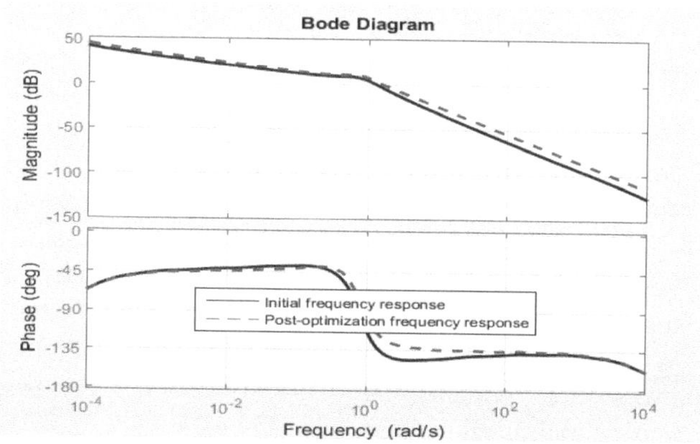

Fig. 4: The Above Plot is the Bode Diagram of Open Loop Transfer Function with Gain Margin of 136.87 dB and with Phase Margin of 59.817°

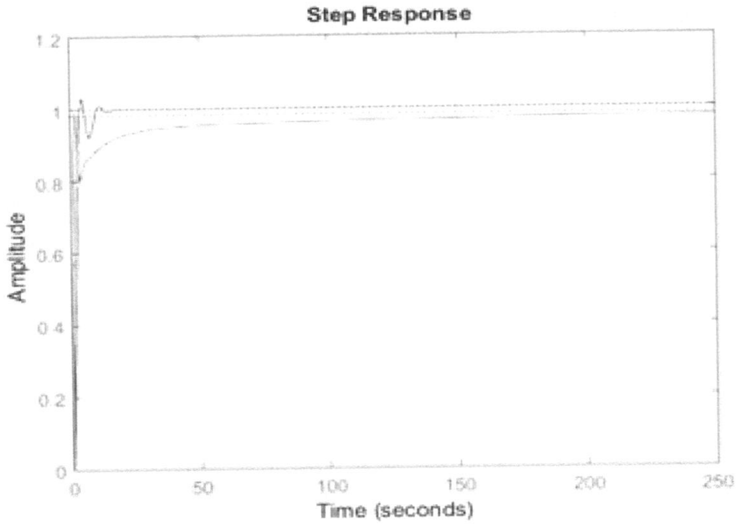

Fig. 5: The above Plot is the Step Response of Closed Loop over All Transfer Function of System, the Graph which is in Red One Corresponds to System with PID Controller and the One in which Green, Corresponds to System with FOPID Controller

Result and Discussion

From the above observations, when a system is controlled with PID controller it will have good performance, instead of it FOPID controller will give better results than PID controller and also performance indices like ISE, IAE, IATE, IASE errors got reduced then accuracy in output is increased. Besides these, time domain specifications improved in value, main parameter over shoot decreased then frequency of oscillations reduced, so overall performance of a system got improved.

The better results are possible with FOPID, because of the two extra parameters. With extra parameters, fractional powers are possible is comparison to traditional PID controller. For example in traditional PID, order of the system can be only integer but in FOPID multiple orders are possible in between any two consecutive integers, so scope of variation trial and error is large, in this process accurate results are possible.

Conclusion

In this paper, the comparison of the PID, FOPID controller mechanisms and their performances and also comparison between different kinds of optimization techniques in order to generate better output. From the Table 9, with Nelder-Mead rise time is less as minimum as 21% and maximum 124%, peak overshoot also less than other methods as minimum 4.6% and maximum 42%. And peak time is less as minimum as 4% and maximum by 22% respectively. These statistics shows that FOPID is superior to traditional PID.

References

[1] Mohd, T.A.T., Hassan, M.K. and WMK. A. Aziz.—Mathematical modelling and simulation of an Electric Vehicle‖. Journal of Mechanical Engineering and Sciences (JMES), Vol. 8, pp. 1312–1321, 2015.

[2] Bova, Nicholas; Marrs, Zachariah; Goodwin, Justin and Oliva, Andrew, Environment and Social Issues Concerned with Hybrid Cars‖, pp. 1–120, 2010.

[3] Wicaksono, Aryo and Prihatmanto, Ary Setijadi. Optimal Control System Design for Electric Vehicle, *4th International Conference on Interactive Digital Media*, Indonesia, 2015.

[4] Mahmud, Md Rasel, Novel Robust Controller Design to Enhance Transient Stability of Power Systems, MS thesis, SWIMBURNE university, Australia, August 2017.

[5] Faisal, Mohammad; Hannan, Mahammad A.; Ker, Pin Jern; Hussain, Aini; Mansor, Muhamad Bin and Blaabjerg, Fred, Review of Energy Storage System Technologies in Microgrid Applications: Issues and Challenges Received April 20, 2018, accepted May 23, 2018, date of publication May 28, 2018, date of current version July 12, 2018.

[6] Guozhen, H.U., ShanxuDuan, Tao C.A.I. and Baoqi, L.I.U., Modeling, Control and Implementation of Lithium-ion-Battery Charger in Electric Vehicle Application, *Electrical Review*, pp. 255–258, 2012.

[7] Farhan A. Salem. Modeling and Control Solutions for Electric Vehicles‖. *European Scientific Journal*, Vol. 9, pp. 221–240, 2013. Tan Zhang, "Adaptive Energy Storage System Control for Microgrid Stability Enhancement" PhD Thesis, WPI university, USA, April-2018. Stefanos Saridakis, Eftichios Koutroulis, Frede Blaabjerg" Optimal Design of Modern Transformerless PV Inverter Topologies", *IEEE Transactions on Energy Conversion,* Vol. 28, No. 2, June 2013.

[8] Hirscha, Adam; Paraga, Yael and Guerrerob, Josep, "A review of technologies, key drivers, and outstanding issues" *Elsevier, Renewable and Sustainable Energy Reviews* 90 (2018), 402–411.

[9] Liang, Chen and Wang, Peng, *Senior Member, IEEE*, Xiaoqing Han, Wenping Qin, Yanbing jia and Tiejiang Yuan Battery Energy Storage Selection Based on a Novel Intermittent Wind Speed Model for Improving Power System Dynamic Reliability, IEEE Transactions 2018, on Smart Grid (volume is at be finalised because it is in editing stage).

Comparative Studies between Phasor Measurement Units (Synchrophasors) and SCADA System

Vinay Hans[1], M.K. Bhaskar[2] and Pappu Ram Bheel[3]

Department of Electrical Engineering, M.B.M. Engineering College, Electrical Engineering Department, Jodhpur, Rajasthan, India

E-mail: [1]vinay.hans1310@gmail.com; [2]mkb31@rediffmail.com; [3]princebhati999@gmail.com

ABSTRACT: *The increasing demand of load without considerable increase in transmission resources has posed numerous constraints and challenges in the power system monitoring and performance. Modern power system grid monitoring tools use data from remote terminal units (RTUs), protective relays, and transducers to provide information to system operators. This information is vital for the operation of the power system on a daily basis and under system contingencies. However, the method presently available to retrieve data from the devices is asynchronous and relatively slow. The asynchronous nature of the data does not provide accurate angle difference information from two nodes on the network. Moreover, the low data rate may be too slow to capture many short-duration disturbances on the grid. Alternatively, Synchrophasors based technology can be used to provide high-speed and coherent real-time information of the power system that is not available from legacy supervisory control and data acquisition (SCADA) systems.*

This paper discusses the existing SCADA system and its comparison with the Phasor Measurement Units. The paper also provides a critical information that is lacking in existing SCADA systems. Authors strongly believe that this survey article will be very much useful to the researchers and power engineers for redefining the existing SCADA systems with the help of Phasor Measurements Units.

Keywords: SCADA; PMU; Real Time Data; Synchrophasors, Smart Grid.

Introduction

Phasor Measurement Unit (PMU) is an instrument that has been developed for improving the protection first and then control and monitoring system in smart power grid. This advanced technology may be used to provide high-speed and coherent real-time information about the power system that is not obtainable in conventional supervisory control and data acquisition (SCADA) systems.

The asynchronous and slow nature of the SCADA system does not provide power system information at subsecond time frames to the state estimator. Therefore, a SCADA/EMS (energy management system) does not provide dynamic state measurement of the power system.

Conventional SCADA system is being used for monitoring and controlling the electric power grids parameters and it receives measurements of grid from Remote terminal units (RTUs) at relatively low rates.

However, Phasor Measurement Units (PMUs) provide real-time data at higher rates (up to 60 times per second) and with higher accuracy. PMUs can also send real-time phase angles directly to SCADA, instead the system estimate the phase angles.

Figure 1 shows a SCADA system performing an asynchronous scan of all the RTUs to retrieve the system data.

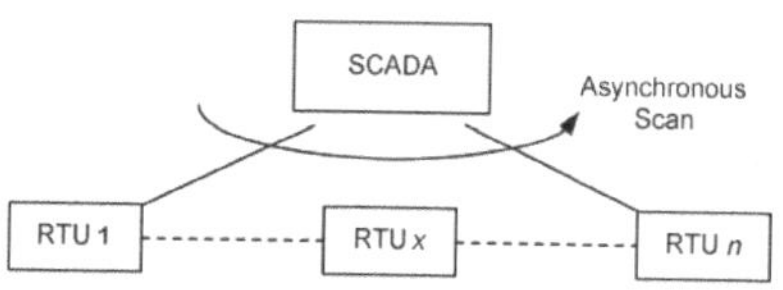

Fig. 1: SCADA System Performing an Asynchronous Scan of RTUs

Figure 2 shows a sinusoidal waveform in time and phasor domain representation, where A is the amplitude of the signal and φ is the phase angle.

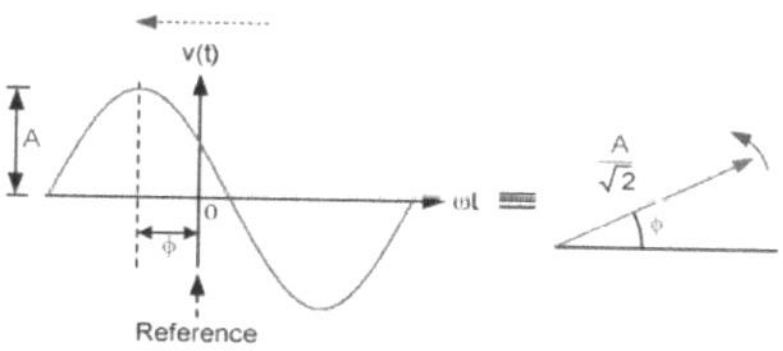

Fig. 2: Phasor Representation of a Sinusoidal Waveform

Overview of SCADA System

Supervisory Control and Data Acquisition or simply SCADA is one of the solutions available for data acquisition, monitor and control systems covering large geographical areas. It refers to the combination of data acquisition and telemetry. SCADA systems are mainly used for the implementation of monitoring and control system of an equipment or a plant in several industries like power plants, oil and gas refining, water and waste control, telecommunications, etc. In this system, measurements are made under field or process level in a plant by number of remote terminal units and then data are transferred to the SCADA central host computer so that more complete process or manufacturing information can be provided remotely.

Conventionally, SCADA systems have made use of the Public Switched Network (PSN) for monitoring purposes. Today many systems are monitored using the infrastructure of the corporate Local Area Network/ Wide Area Network (LAN)/(WAN).

A SCADA/EMS supervises, controls, optimizes, and manages generation, transmission, and distribution assets. Typical functionality of a SCADA/EMS includes.

- Data acquisition from RTUs and protective relays.
- Postprocessing of data for sanity checks and scaling.
- Data archiving for postmortem analysis.
- Sequence-of-event recording.
- State estimation.
- Manual and automatic generation control.
- Load forecasting.
- Contingency analysis.
- Offline tools for power flow, contingency, and scheduling.
- Operator interface for visualization and additional programming for custom applications.

Phasor Measurement Unit (PMU)

Phasor measurement units are devices which measures frequency, magnitude and phase angle of voltage and current with respect to a time reference. Unlike the conventional measuring instruments, PMUs measure the analog waveforms and provide digitized output, with GPS reference source and accuracy of 1 microsecond. These high precision synchrophasor data composed of both magnitudes and phase angle, are transmitted to the remote servers, local substations. Effective utilization of this technology is very useful in mitigating blackouts and learning the real time behavior of the power system. This drastically improves visibility and expedites truthful fault diagnosis..

The synchronized phasor measurement technology is relatively new, and consequently several research groups around the world are actively developing applications of this technology. It seems clear that many of these applications can be conveniently grouped as follows:

1. Power System Real Time Monitoring
2. Advanced network protection
3. Advanced control scheme.

PMUs facilitate innovative solutions to traditional utility problems and offer power system engineers a whole range of potential benefits, including:

- Post-disturbance analyses are much improved because precise snapshots of the system states are obtained through GPS synchronization.
- Precise estimates of the power system state can be obtained at frequent intervals, enabling dynamic phenomena to be observed from a central location, and appropriate control actions taken.
- To analyze the vulnerability of the system against any contingency. This is known as security assessment of the power system networks.
- Advanced protection based upon synchronized phasor measurements could be implemented, with options for improving overall system response to catastrophic events.
- To ensure acceptable quality of the power supplied to the consumers.
- Advanced control using remote feedback becomes possible, thereby improving controller performance.

The Complete block diagram of PMUs and PMUs utilization in power systems are shown in Figure 3.

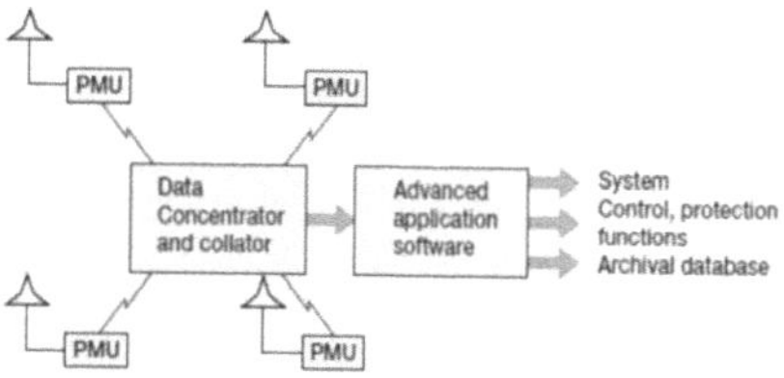

Fig. 3: PMU Utilization in Power System

The various features of PMUs are given below as follows:

- PMUs are Measures 50/60 Hz AC waveforms (voltage and current) typically at a rate of 48 samples per cycle.
- PMUs are then computed using Discrete Fourier Transform (DFT)-like algorithms, and time stamped with a GPS.
- The resultant time tagged PMUs can be transmitted to a local or remote receiver at rates up to 60 samples per cycle.

The application of PMUs to perform controlled islanding in power systems that are lightly meshed/radial has shown to work well. However, in heavily meshed power systems, no production-grade real-time applications have been implemented yet. A controlled system separation using PMU measurements can greatly improve the performance of such a scheme. The implementation requirements depend on the type and complexity of the scheme and the role of PMU measurements. A large number of PMUs need to be installed to realize a system-wide islanding scheme. These PMUs need a communications infrastructure to support the large amount real-time data transfer. Synchrophasors are well suited for steady-state and quasi steady-state phenomena (not for transient conditions like faults) and for observing low-level oscillations.

There are a number of algorithms for coherency detection and self-sufficient island identification, but further improvements are still needed. Using real-time information from PMU and automated controls to anticipate, detect, and respond to system problems, a smart grid can automatically avoid or mitigate power outages, power quality problems, and service disruptions. A phasor network consists of PMUs installed in the power system buses, PDC to collect the information and SCADA system at the central control facility as shown in Figure 4.

Such a network is used in Wide area measurement system (WAMS).

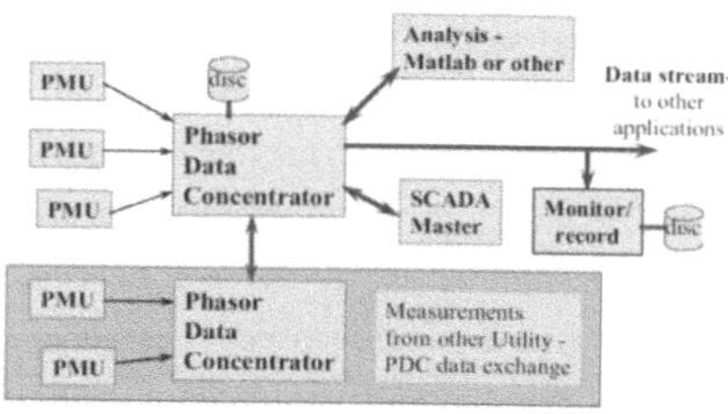

Fig. 4: Phasor Measurement System

Comparative Studies between RTU and PMU based SCADA

State estimators are commonly used to estimate the state of the power system;

power system operators use this information to make decisions about operating the power system under normal and contingency conditions. The front end of a state estimator is the supervisory control and data acquisition (SCADA) system that gathers system data, which include voltage magnitudes, active and reactive power values, and system topology via breaker status. The state estimator uses this information and system network impedances to estimate the state of the power system. The SCADA system periodically polls the devices that collect the information, which can include remote terminal units (RTUs), protective relays, and transducers, to obtain the data. One complete scan of the large number of devices could last from 2 to 10 seconds. During normal steady-state conditions, the long scan time is not a major concern. However, when the system state changes during the scan, the data retrieved no longer represent 56 the system state accurately. The asynchronous and slow nature of the SCADA system does not provide power system information at subsecond time frames to the state estimator. Therefore, a SCADA/EMS (energy management system) does not provide dynamic state measurement of the power system.

Synchrophasors provide phasor measurements of voltages and currents with a common time reference; the GPS time source is one such common time reference. With synchrophasors, the state of the power system can be measured, as compared with state estimation using a traditional SCADA.

Phasor measurement units (PMUs) sample voltage and current many times a second and accurately time-stamp each sample. This technology can be used to provide high-speed and coherent real-time information of the power system that is not available from legacy supervisory control and data acquisition (SCADA) systems.

PMUs connected to a GPS satellite clock send voltage and current phasor measurements to a phasor data concentrator (PDC) at a rate of 25 messages per second, for example. The PDC receives data from multiple PMUs, time aligns the data, and provides the super packet to upstream applications. A super packet consists of phasor measurements from different PMUs with a common time stamp.

Synchrophasors are being used in SCADA by many utilities. The key objective of making synchrophasors available to SCADA is to enhance state estimation. Synchrophasors provide high-speed (subsecond) coherent data that are not available with traditional SCADA measurements in order to monitor power system dynamics.

Figure 5 shows the building blocks of a Phasor Measurement Unit-based system.

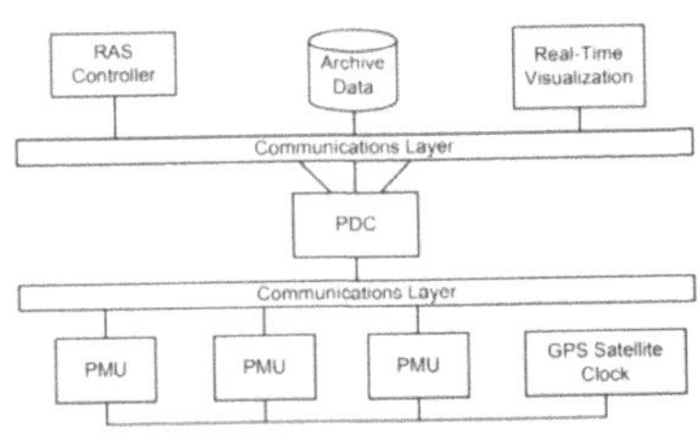

Fig. 5: Phasor Measurement Unit based System

Table 1: Comparisons between RTU and PMU based SCADA

Attributes	*RTU based SCADA*	*PMU based SCADA*
Measurement	Analog	Digital
Resolution	2–4 samples per cycle	Upto 60 samples per cycle
Observability	Steady State	Dynamic/Transient
Monitoring	Local	Wide Area
Phasor Angle Measurement	No	Yes

The comparisons between RTU and PMU based SCADA systems are given in Table 1. It is concluded that the PMU gives more accurate and efficient results than RTU in SCADA.

Conclusion

It is very hard to find out the cause of blackout and determine the root cause of large-scale disturbance. When recorded data is not real time synchronized, so PMU is solution for this problem. PMU provides means for accurate long-term wide-area system dynamic recordings. The high data rate and precision allow this technology to capture fast changing system dynamics of the power system.

The WAMS proved to be beneficial in the analysis of the events and provided details of the system dynamics that were previously hidden because of the asynchronous and slow data rate of the SCADA system. The measurements taken by a wide-area system dynamics recording technology are also helpful in determining power system characteristics such as line parameters, thermal limits, load models, machine performance, and parameters of its associated control system models. All real time synchronized data can be recorded.

Asynchronous polling of the SCADA measurements resulted in different frequency measurements. This difference does not exist in the power system, as demonstrated by the synchrophasor frequency measurements.

Frequency and df/dt measurements in the synchrophasor message were used to validate the operation of under-frequency and df/dt relays. Synchrophasors provide information of low-level oscillations; tools are available to calculate oscillation frequency and damping ratio in real time. Synchrophasors based PMU system provide voltage phase angle difference information that is critical for power system operators to monitor the stability of the power system.

References

[1] Bentarzi, Hamid and Tsebia, Mohamed, "PMU Based SCADA Enhancement in Smart Power Grid", *IEEE (CPE-POWERENG 2018)*, April 2018.

[2] Mynam, Mangapathirao V.; Harikrishna, Ala and Singh, Vivek, "Synchrophasors Redefining SCADA

Systems", *13th Annual Western Power Delivery Automation Conference*, March 2011.

[3] Novosel, Damir and Vu, Khoi, "Benefits of PMU Technology for various applications", *Published in International Council of Large Electric Systems, 7th Symposium on Power System Management*, November 5th–8th, 2006.

[4] Rahman, Waheed Ur; Ali, Muhammad; Mehmood, Chaudhry A. and Khan, Asadullah, "Design and Implementation for Wide Area Power System Monitoring and Protection using Phasor measuring Units", *Published in WSEAS Transaction on Power Systems*, Vol. 8, Issue 2, April 2013

[5] Singh, Bindeshwar; Sharma, N.K.; Tiwari, A.N.; Verma, K.S. and Singh, S.N., Applications of phasor measurement units (PMUs) in electric power system networks incorporated with FACTS controllers, *Published in International Journal of Engineering, Science and Technology*, Vol. 3, Issue 3, pp. 64–82, 2011.

[6] Saadat, Hadi (2002). Power System Analysis, 2nd ed., New York: Mc Graw-Hill.

Author Index